J.S. Mill Revisited

J.S. Mill Revisited

Biographical and Political Explorations

Bruce L. Kinzer

J.S. MILL REVISITED
Copyright © Bruce L. Kinzer, 2007.

All rights reserved. No part of this book may be used or reproduced in any manner whatsoever without written permission except in the case of brief quotations embodied in critical articles or reviews.

First published in 2007 by
PALGRAVE MACMILLAN™
175 Fifth Avenue, New York, N.Y. 10010 and
Houndmills, Basingstoke, Hampshire, England RG21 6XS
Companies and representatives throughout the world.

PALGRAVE MACMILLAN is the global academic imprint of the Palgrave Macmillan division of St. Martin's Press, LLC and of Palgrave Macmillan Ltd. Macmillan® is a registered trademark in the United States, United Kingdom and other countries. Palgrave is a registered trademark in the European Union and other countries.

ISBN-13: 978–1–4039–8068–7
ISBN-10: 1–4039–8068–3

Library of Congress Cataloging-in-Publication Data

Kinzer, Bruce L., 1948–
  J.S. Mill revisited : biographical and political explorations / Bruce L. Kinzer.
    p. cm.
  Includes index.
  ISBN 1–4039–8068–3 (alk. paper)
   1. Mill, John Stuart, 1806–1873. 2. Political scientists—Great Britain—Biography. 3. Philosophers—England—Biography. I. Title.

JC223.M66K56 2007
320.092—dc22   [B] 2007009018

A catalogue record for this book is available from the British Library.

Design by Newgen Imaging Systems (P) Ltd., Chennai, India.

First Edition: October 2007

10 9 8 7 6 5 4 3 2 1

Printed in the United States of America.

*For Deborah, Anna, Amanda,
Jon, Kian, and Lara*

CONTENTS

Acknowledgments		viii
Introduction		1
One	The Father, the Son, and the Manly Spirit	8
Two	Gathering Truths, 1826–30	43
Three	Mill and Harriet Taylor: The Early Years	77
Four	Mystifying the Mystic: Mill and Carlyle in the 1830s	112
Five	Mill and the Secret Ballot	146
Six	Mill and the Problem of Party	164
Seven	Mill and the Experience of Political Engagement	179
Notes		205
Index		232

ACKNOWLEDGMENTS

Earlier versions of three of the chapters in this book appeared in print under the following titles and in the following places: "J.S. Mill and the Secret Ballot," *Historical Reflections/Réflexions Historiques* 5 (1978): 19–39; "J.S. Mill and the Problem of Party," *Journal of British Studies* 21 (1981): 106–22; "John Stuart Mill and the Experience of Political Engagement," in *A Cultivated Mind: Essays on J.S. Mill Presented to John M. Robson*, edited by Michael Laine (Toronto: University of Toronto Press, 1991), 182–214. I am grateful to the editors, journals, and publishers concerned for permission to reprint these items in a revised form. I am also grateful to the University of Toronto Press for allowing me to quote liberally from the great Toronto edition of *The Collected Works of John Stuart Mill* (1963–91). Mill scholars everywhere owe an enormous debt to John M. Robson, the general editor of the *Collected Works*, and none can possibly owe more than I. The chapter on Mill and Carlyle could not have been written without use of the superb Duke University Press edition of *The Collected Letters of Thomas and Jane Welsh Carlyle* (general editor, Charles Richard Sanders). For permission to quote extensively from this edition, I wish to thank Duke University Press. I am also grateful to Indiana University Press for permission to quote at some length from *The Complete Works of Harriet Taylor Mill*, edited by Jo Ellen Jacobs (1998). I want to express my appreciation to the National Art Library, Victoria and Albert Museum, for allowing me to quote from the diary of Henry Cole, material reproduced with the kind permission of the Trustees of the Victoria and Albert Museum.

One person especially, my friend Jeff Lipkes, had a close association with the four biographical chapters in this book. He read each with great care and offered much penetrating criticism (along with words of comfort). I am grateful for the use to which he applied his acute intellect and impressive knowledge of Mill, and more grateful for his friendship.

Martin Sheppard also read early versions of these chapters, and provided excellent editorial advice. I asked my wife, Deborah Kinzer, to read a draft introduction to this book, and her astute comments produced changes in tone that I can now see were sorely needed. William McCulloh and Patricia McCulloh kindly agreed to read the piece on Mill and Carlyle, and their observations gave rise to a number of improvements. I thank them. As I do my dear friend Ann Robson, who made available to me valuable material on Mill's activity compiled by members of the Mill Project in days past. Most of the writing done for the biographical chapters was carried out during the 2004–05 academic year, when I was on leave. The generous sympathy of my colleagues in Kenyon's Department of History, together with the gracious support of the Provost of the College, made this leave possible. A version of the first chapter in this book was presented to the Kenyon Seminar, whose members furnished both welcome encouragement and stimulating critical commentary. For mistakes in this book, of fact or judgment, I am alone responsible.

As for the crowded character of the dedication page, the persons named are immeasurably precious to me, and an opportunity of this kind may not come again.

Introduction

This book consists of two groups of chapters. The first of these, biographical in character and of recent composition, offers a reconsideration of J.S. Mill's formative period. The second, whose provenance dates back some years, examines topics and themes bearing upon Mill's political thought and activity. Intellectual authority mattered to Mill principally because of the moral and political ends it could be made to serve. The latter set of chapters, in sundry ways, explores the relation between some of these ends and Mill's conception of himself as a public intellectual. The biographical chapters investigate challenges he faced in trying to sort out what he thought of himself and what he wished to become. In part they take the form they do because of what I have become, and the challenges I have faced in trying to sort out what I think of J.S. Mill.

I have visited with Mill's politics for a long time, and have deliberately kept to a minimum excursions that involved tackling aspects of Mill's life and works that did not relate directly to problems associated with his political thought and activity. Trained as a political historian, I had a fervent interest in Victorian political history. My turn to Mill scholarship was fortuitous and happened only after I finished my graduate work at the University of Toronto. I audited a course on Victorian prose taught by John M. Robson, general editor of Mill's *Collected Works*. In luck's grip, I soon found myself in the enviable position of postdoctoral fellow on the Mill Project. With a wealth of stimulating material at hand, and the hearty and generous encouragement of the editor of the *Collected Works*, I began to look at circumscribed spheres of Mill's political engagement. The probing continued for the next quarter century.

I perforce learned a fair amount about Mill that did not impinge in a straightforward way on the political issues I grappled with. If anything, this learning made me even more wary of confronting problems of a biographical nature that held a controversial place in Mill studies: the psychological impact of James Mill on his eldest son; the causes and nature of J.S. Mill's "mental crisis"; the dynamics of the Mill-Harriet Taylor relationship. Although I appreciated that any rounded understanding of Mill depended on coming to grips with these problems, I was generally content to concentrate on narrow political matters that I thought had been insufficiently explored in the existing literature. In doing so, I recognized that I could thereby elucidate only a particular side of Mill, but it was a side that keenly interested me.

A few years ago an opportunity arose to step away from the safe niche I had nestled into. Encouraged to embark upon a biographical investigation of Mill's early decades, with a view to sizing up his development and character, I reckoned it was time to take a risk. I had reached the twilight of my career without having weighed in on Mill's formative years. The time had come to revisit this period of Mill's life in a systematic and purposeful way, so that I might work out what I thought of the young person who ultimately became the preeminent thinker of his generation.

My approach involved an effort to shield myself from the large body of excellent secondary literature on Mill. I knew full well that I had, over the years, amassed an incalculable debt to the many Mill scholars who had enriched my understanding of his life and thought. I now considered it necessary, if ungrateful, to put their work aside. While I knew I could never bring a completely open mind to my revisiting of Mill's early years, I wanted to keep the preconceptions to a minimum. Had I begun this undertaking with decided convictions of my own about what I wanted to say, I would not have felt a need to maintain a certain distance from the formidable gathering of Mill scholarship that has taken shape over the last half century. My aim was to see what I could learn about the young John Mill through a largely unmediated encounter with the relevant primary materials.

The method I embraced also urged me to resist using what I thought I knew about the mature J.S. Mill to understand and explain the young John Mill. Convinced that there was something to be gained from looking at his first three decades in isolation from what he later became, I sought to come to terms with important aspects of Mill's formative years mainly without reference to his subsequent activities and achievements. I think this inquiry has yielded some fruitful insights, and hope that others will agree, including those interested in what Mill said and did—and did not say or do—after his thirtieth birthday.

Although the opening chapter does not tell a story about Mill's youth and education fundamentally at odds with that told by Michael St. John Packe and Nicholas Capaldi in their full-scale biographies, it does emphasize certain themes not prominently featured in these valuable works.[1] While acknowledging the many good things Mill derived from the education he received from his father, I suggest that these good things could have been acquired in conditions much less severe than those James Mill imposed. I also stress the importance of what John Mill refers to as his "habitual reserve," and his "instinct of closeness," characteristics inseparable from the strains of his upbringing.[2] Holding no brief for the psychoanalytical diagnoses of Mill's youth offered by Bruce Mazlish and P.J. Glassman, I yet accept that James Mill, however inadvertently, hampered his son's development of a coherent identity.[3] At the same time, I contend that if the father's impact had a lot to do with the son's mental crisis, the measure of success John Mill had in overcoming this crisis also owed something to his father. In addition to these points, I make more than others have made of John Mill's first experience of feeling "a man among men" when in the company of Charles Austin and his friends.[4]

The three biographical chapters that follow have a more concentrated temporal focus than the first. One covers the years from 1826 to 1830, during which Mill—then in his early twenties—continued to absorb the effects of the mental crisis. Mill met Harriet Taylor in 1830, and the third chapter in the book examines the early years of the Mill-Harriet Taylor relationship. Not long after his connection with Harriet Taylor began, Mill came into contact with Thomas Carlyle, and the last of the biographical pieces examines the interaction of Mill and Carlyle in the first half of the 1830s. Like the first chapter, the three that come next chiefly interest themselves in matters pertaining to Mill's character. Although each of the four biographical chapters can stand alone, the first two set the stage for what follows by conveying my notion of who John Mill was on the eve of his encounters with Taylor and Carlyle. In the latter two chapters I try to discover what Mill's friendship with each of these powerful personalities tells us about his character.

The subject of Mill's character is at the center of Janice Carlisle's brilliant and ambitious *John Stuart Mill and the Writing of Character* (1991). The dimensions of Carlisle's canvas encompass the entire span of Mill's life, and therefore extend far beyond the scope of my tableau. Her thesis, however, hinges on her interpretation of Mill's formative years. Her intricate chain of reasoning incorporates the following links. Mill's early education and rigorous grounding in the psychology of association instilled in him a conviction that his character was embodied in his writings. His famous crisis had more to do with vocation than emotion. Above all else, he wanted for himself a parliamentary career that would satisfy his longing to be a practical—and heroic—reformer of the world. The onset of the crisis, Carlisle notes, coincided with the completion of his apprenticeship at India House and his consequent recognition that a parliamentary career was out of reach. Mill's occupational constraints and confinement within the sphere of theory bore a heavy psychic cost as he struggled to assimilate the marginal status to which he felt he had been reduced. The effects of this struggle moved Mill to direct his message to an audience of the future. Inasmuch as he could not participate practically in reforming the world in which he lived, the reforms with which he came to identify had to be the work of future generations that would have the wisdom to value his work as his contemporaries could not. The "friendship" with Harriet Taylor signified and rigidified Mill's marginality, their way of life distancing Mill further from the realm of practical political action and underscoring his "continued subjection to external forces."[5] The limitations and disappointments he experienced forcefully affected Mill's writings.

I know that my reaction to Carlisle cannot be detached from my perception of who the mature Mill was, just as I know that this perception inevitably colors my understanding of who he was before he became an eminent Victorian. Readers of this book will allow for the influence of this distorting lens when I give my opinion that Mill never conceived of himself as a marginal figure in relation to the culture of which he was a part. In his early manhood he was no doubt sometimes unsure about the precise place he would be able to claim for himself within this culture. What young man of twenty years, Mill's age when he completed his apprenticeship with the East India Company, feels certainty about such a matter? Yes, he regretted that his position at India House stood in the way of his ambition to sit in the House of Commons. He did not, in my view, see that it stood in the way of his being a practical reformer. John Mill regarded his father, a servant of the East India Company for the last seventeen years of his life, as an astute and notably effective man of action whose impact on his age was far from negligible. Employment at India House did not keep either James Mill or his son from giving practical political expression to their radicalism. In the chapter on J.S. Mill in the late 1820s I hold that he was well-situated for pursuing the aims he defined for himself in the wake of the mental crisis. What personal confusion, insecurity, and loneliness he experienced in these years did not arise from a sense of thwarted ambition or cultural insignificance.

Carlisle's commentary on the effects of Mill's intimacy with Harriet Taylor chiefly relate to the period after that explored in the set of biographical chapters presented in this book. Much has been written on the Mill-Taylor friendship. The pioneering modern work on the Mill-Taylor association is F.A. Hayek's *John Stuart Mill and Harriet Taylor: Their Friendship and Subsequent Marriage* (1951). Hayek acts more as editor than author, and devotes two fairly short chapters to the early years of the friendship. One of these consists principally of the relevant surviving correspondence, with Hayek providing informational passages by way of introduction and conclusion. Editorial interpolations within the chapter take the form of brief commentaries that serve to link the documents. The other chapter prints the essays "On Marriage and Divorce" that Mill and Taylor exchanged early in their relationship.[6] Packe's biography appeared three years after the publication of Hayek's volume. Despite Packe's tendency to let his imagination run away with him, his account of the early years of the Mill-Taylor friendship offers a strong narrative line.[7] It says very little, however, about the essays "On Marriage and Divorce," from which, I believe, much can be learned about the respective parties as they were in the early 1830s. Capaldi, who by no means understates the significance of Taylor's influence on Mill's intellectual priorities, does not allot much space to the early years of the friendship.[8] Nor can a great deal be learned about these years from Josephine Kamm's *John Stuart Mill in Love* (1977), a readable but flawed work that considers the effects of a number of women on Mill (Harriet Taylor, of course, but also his mother, Sarah Austin, Eliza Flower, Harriet Martineau, and Helen Taylor).[9] The book's purpose does not allow for an in depth treatment of the early phase of the Mill-Taylor connection. Similarly, Phyllis Rose's suggestive essay on the Mill-Taylor relationship in her book, *Parallel Lives: Five Victorian Marriages* (1983), cannot accommodate a detailed investigation of its early years.[10] These years do receive liberal attention in a recent work by Jo Ellen Jacobs, *The Voice of Harriet Taylor Mill* (2002).[11] Yet the unorthodox nature of this attention—a fictitious Harriet Taylor diary created by Jacobs—renders hazardous any comparison of my findings with those of Jacobs (especially since Mill occupies a subordinate place in her account, while Taylor occupies a subordinate place in mine). The question of Harriet Taylor's influence on Mill has given rise to a considerable literature, most of it concerned with the character and extent of Taylor's impact on what Mill wrote between 1848 and 1858.[12] Although my discussion closes with a brief assessment of Harriet Taylor's influence on Mill's intellectual development during the 1830s, it has nothing to say about what happened thereafter. A good deal has also been written on Mill and the issue of sexual equality.[13] My look at the early years of the Mill-Taylor friendship may indirectly have a bearing on this problem, one that is incidental to the object of the essay. The limited scope and narrowly biographical purpose of the chapter afford special analytical opportunities, which I have tried to seize.

Scholars pay attention to Harriet Taylor mainly because of her importance in Mill's life; Carlyle's historical standing in no way depends on his association with Mill. While anyone trying to write a comprehensive biography of either man would certainly have occasion to say something about the other, none would feel called upon to furnish a comprehensive account of their friendship during the 1830s. Packe has more to say on the matter than Capaldi, but neither attempts a systematic and detailed investigation of the subject (such a treatment would surely have affected adversely the rhythm and pace of their studies). The same can be said of Fred Kaplan's biography of Carlyle.[14] A couple of books have been devoted to a comparative examination of the thought of Mill and Carlyle. In these works the two men serve as representative figures of their age. The titles aptly convey the purview of each book. Emery Neff's major scholarly study, *Carlyle and Mill: Mystic and Utilitarian*, was published in 1924; it was republished in 1926, with a new subtitle, *An Introduction to Victorian Thought*.[15] Richard Pankhurst's less ambitious (and less satisfying) work, *The Saint Simonians, Mill and Carlyle: A Preface to Modern Thought*, appeared in 1957.[16] Both books, of course, discuss Mill and Carlyle in the 1830s, but with ends in mind that differ from the aim of my chapter. This aim is not to explore the thought of Mill and Carlyle during the thirties, but to learn what their friendship reveals about Mill's understanding of himself.

The last third of *J.S. Mill Revisited* consists of three chapters on political themes. Two of these initially took the form of journal articles; the last of the three I composed for a festschrift honoring John Robson. I hope my bringing them together as a set will be welcomed by readers interested in Mill's politics. Although I have rewritten the essays on the ballot and on party to render the prose more readable, their substance remains essentially unchanged. The Mill scholarship on the topics themselves, scanty before I tried to explain his change of mind on the ballot and his notion of party, has not grown much since the journal articles appeared. The index to *The Cambridge Companion to Mill*, an estimable collaborative work that runs to nearly six hundred pages, has no entries for the ballot or for party.[17] Among Mill scholars, to be sure, philosophers greatly outnumber historians, and the priorities of the former will not steer them toward Mill's apostasy on the ballot question or his attitude toward party. I readily admit that there is not a great deal to be said on either subject; there is, however, something to be said (including, no doubt, something different from what I have chosen to say). Few political historians, as distinct from historians of political thought, have seen fit to make Mill their principal object of study. I happened to be drawn to these particular problems because of my abiding interest in Victorian politics. The treatment I give them is in keeping with my belief that a heavy dose of contextualization can enhance our understanding of Mill's political thinking. The chapter capping this book, which investigates the ebb and flow of Mill's political radicalism between the 1830s and the late 1860s, draws on this conviction to illuminate the apparent disparity between Mill's political theory and his political activism. Its fusing of biographical and political elements makes it a suitable finale.

CHAPTER ONE

The Father, the Son, and the Manly Spirit

It is not easy to discuss J.S. Mill's childhood and education without regard for what he ultimately became. The principal source for that childhood and education is his *Autobiography*. The first draft of this classic account was composed in 1853–54, when Mill was middle-aged and justly renowned as the author of *A System of Logic* (1843) and *The Principles of Political Economy* (1848). Its purposes were shaped by the personal and didactic concerns of Mill and his wife, Harriet Taylor Mill, during the first half of the 1850s. In the opening paragraph of what has come to be known as the "Early Draft," Mill offers two reasons for writing his autobiography: first, the age in which he lived evinced a conspicuous interest in education and a record of his own "unusual and remarkable" training could throw valuable light on what could be learned in childhood; second, that age was also one of "transition in opinions" and as such stood to gain from "noting the successive phases of any mind which was always pressing forward, equally ready to learn and unlearn either from its own thoughts or from those of others." The final version of Mill's *Autobiography* (he last worked on the text in 1869–70) adds a third object, one that mattered to him, he says, more than the others. He wanted to pay formal acknowledgment to those who contributed to his moral and intellectual growth, "some of them of recognized eminence, others less known than they deserve to be, and the one to whom most of all is due, one whom the world had no opportunity of knowing."[1] Harriet Taylor Mill was the person "to whom most of all is due."

A fourth aim, though no less fundamental than the other three, is left unstated in the opening paragraph. Mill and Taylor wished to seize control over the telling of their own story. In Taylor's words, "every ground should be occupied by ourselves on our own subject."[2] Only the parties to the relationship had the means (and the motive) to show the purity and high-mindedness that continuously informed it.

As a source of information on Mill's early years, the *Autobiography* is indispensable; as a source of judgments about these years, it must be treated with caution. All three of his stated purposes relate in some way to his development. The perspective he adopts on his formative years is markedly influenced by his conception of what he had become and the means by which he had got there. His explicit motives, expressed in plain and concise language, overlay a deep and rich subsoil of intention. In giving effect to purposes one and two, Mill meant to show, among other things, that the doctrine of Philosophical Necessity—that is, "formation of character by circumstance"—enjoyed a more limited writ than was often supposed. Mill tells his readers that this doctrine, at a critical juncture in his life, had "weighed on my existence like an incubus." The outcome of his intellectual and psychological struggle to lift this burden was a conviction that the human will could influence circumstances and thereby affect subsequent motives and actions.[3] Notwithstanding an education that had greatly privileged analysis over feeling, Mill found within himself the resources to cultivate his imagination and sympathies. He achieved a balance that allowed him to attain both mental and emotional maturity. This achievement also helped him to take measure of the currents and cross-currents of opinion symptomatic of an age of transition. Drawing upon his training and his will, he zealously sought to construct a coherent synthesis, the *Autobiography* registering in summary form the learning, unlearning, and better learning this process entailed.[4] In these ways Mill uses his personal story to exemplify the triumph of will and intellect necessary to grasp the spirit of the age and to promote human improvement. His treatment of his childhood is governed in part by his sense of its bearing upon the problem of determinism and his resolution of that problem; in part by his need to link the early years of his life to the subsequent stages of development that made him a representative figure of his times and an individual capable of making a profound impact on those times; and in very considerable part by his need to raise the ghost of his father, James Mill, and lay that ghost to rest.

The forceful personality and imposing intellect of James Mill pervade the opening chapters of the *Autobiography*. In stating that his supreme motive for writing his story was to acknowledge what he owed to others, the younger Mill had his father preeminently in mind when he referred to "others less known than they deserve to be." Far less well-known than James Mill, at the time J.S. Mill told his story, was Harriet Taylor, for whose sake much of what he had to tell needed to be told. J.S. Mill would have his readers think that he owed less to James Mill than to Harriet Taylor. No discerning reader of the *Autobiography* is likely to accord Harriet Taylor this precedence. Mill's absurdly inflated encomium has the effect of making Harriet Taylor come across as the fictive creation of an author who had permitted his emotional needs to corrupt his judgment. The James Mill of whom we read in the *Autobiography*, in contrast, has a compelling presence whose sway was unmistakably colossal. Yet he could not be given top billing without compromising the standing his son was bent upon assigning Taylor. Equally significant, recognition of James Mill as the paramount influence in J.S. Mill's mental and moral development was incompatible with the spin the latter needed to give the doctrine of Necessity in relation to his own story. If J.S. Mill was indeed a "manufactured man," as he says he at one time feared, the identity of his maker could not be doubted. The younger Mill had to rid himself of the "incubus" and show that he had been able to take a decisive hand in forging his own development. In paying tribute to his father he had also to illumine the path by which he won independence from him. Harriet Taylor did not present anything like the same problem. She was Mill's contemporary, not his senior; she had nothing like the education he possessed; she did not threaten his idea of how he got to be the man he was. Mill had handled the most serious effects of the "crisis" in his "mental history" before he met Harriet Taylor in 1830. Part of what made him her worthy partner, we are led to assume, was the independent growth he had experienced in the several years after the onset of the crisis. The tidy structure J.S. Mill wanted to give his development could not accommodate a prominent role for his father in the years after 1826. In the *Autobiography* he strove to give a fair assessment of his father's faults and virtues as a thinker, educator, and man; it was far more difficult for him to come to terms with how much he remained his father's son even after attaining manhood.[5] To appreciate this difficulty we must investigate the years before the younger Mill reached maturity.

★ ★ ★

The man who molded John Mill's early moral and intellectual growth was born in the Scottish hamlet of Northwater Bridge, Forfarshire, in 1773.

James Mill was the eldest son of James Milne, a shoemaker, and Isabel Fenton. Alexander Bain, James Mill's biographer, found that local tradition remembered Isabel Fenton as a woman of notable social pretension.[6] Her son's certificate of birth gave his surname as "Mill," a form more English than Scottish. As a child James Mill showed signs of unusual ability, and his mother aspired to see her talented son rise to a position of consequence. Given ample opportunity for study despite his family's lack of resources, James Mill did not squander his main chance. Academic success at the local parish school won him the solicitude and encouragement of the local minister, and the means were found to send him to Montrose Academy. Having made a name for himself as an exceptionally gifted young scholar, he was taken up by Sir John and Lady Jane Stuart, who spent their summers not far from James Mill's birthplace. They employed Mill to act as tutor to their daughter Wilhelmina, for whom James Mill developed powerful feelings. Just how powerful can be seen in a letter he wrote to his friend Francis Place in 1817, years after Wilhelmina's death. "So you see I owe much to Sir John Stuart, who had a daughter, one only child, about the same age as myself, who besides being a beautiful woman, was in point of intellect and disposition one of the most perfect human beings I have ever known. We grew up together and studied together from children, and were about the best friends that either of us ever had."[7] The Stuarts could not regard the son of a shoemaker as a suitable candidate for their daughter's hand; they could see their way to sponsoring James Mill's matriculation at the University of Edinburgh, in the expectation that he would study for the ministry. James Mill dutifully complied, and then obtained a license to preach. It seems he made sparse use of this qualification. In 1802 Mill, like many Scots before him, and not a few after, left for London, where he intended to make his living by his pen.

Make his living he did, gaining sufficient employment as journalist and editor of ephemeral weeklies and monthlies to consider marriage within less than three years after his arrival in London. In 1804 he proposed to Harriet Burrow, whose mother, a widow, operated an asylum for lunatics. They wed the following year. Less than a year later, Harriet Mill gave birth to a son. When this son, John Stuart, composed the final draft of his *Autobiography* he did not mention his mother. He began the account of his life with the sentence: "I was born in London, on the 20th of May 1806, and was the eldest son of James Mill, the author of The History of British India."[8] To father and son alike, she evidently was something of an embarrassment. It is safe to assume that it was not the quality of her mind that had drawn James Mill to her. Bain notes that she was "an exceedingly pretty woman."[9] Twenty-two years of age at the time of her marriage—James Mill was then thirty-two—the bride could have had but little idea of what lay in store for her. As Bain puts it: "Mrs. Mill was not wanting in any of the domestic virtues of an English mother. She toiled hard for her house and children, and became thoroughly obedient to her lord." This obedience did not win her James Mill's continuing affection and regard; neither did it make Harriet Mill contented with her lot. "An admired beauty," Bain observes, "she seems to have been chagrined at the discovery of her position after marriage. There was disappointment on both sides: the union was never happy."[10] On James Mill's side the disappointment was almost certainly compounded by fond memories of Wilhelmina Stuart.

Although unhappy, the marriage had a copious aspect. John would be the eldest of nine children. Between his birth in May 1806 and his father's India House appointment as an Assistant to the Examiner of India Correspondence in May 1819, James Mill's rapidly expanding family threatened to outstrip his limited means. Not long after John's arrival, his father set to work on the history of British India, a project he calculated would take three years to complete. It took ten years, during which James Mill's modest earnings derived chiefly from articles he contributed to the *Edinburgh Review*, the great Whig periodical. In these years James Mill also took it upon himself to educate his eldest son, which presumably further reduced the time available for producing ephemeral pieces that would have brought in additional income. Jeremy Bentham, during this trying stretch, made a philanthropic project of the Mills. James Mill's struggle to keep out of debt proved somewhat more successful than his struggle to keep out of his wife's bed. He understood, on personal economic grounds and on general Malthusian lines, the importance of sexual restraint. An impressive self-mastery figured among the attributes recommending the elder Mill to those who greatly admired him. In his conjugal relations, however, self-mastery eluded him. James Mill's contempt for his wife may have arisen not only from what he took to be her deficient intellect; he may also have held her responsible for his inability to control a physical appetite that gave him more children than he knew what to do with. He thought he knew what to do with John, and when the others came along they were turned over to the eldest child for their lessons. For both James Mill and his first-born, Mrs. Mill's fecundity was something of an affliction.

This is not to suggest that the young John Mill passed his time thinking about the regrettable results of his mother's reproductive performance. His education, which began early, was wholly under the direction of his father. James Mill's aims, method, and manner ensured that he would be at the center of the boy's universe. The elder Mill's belief in the cardinal importance of education, coupled with his conviction that no school in early nineteenth-century Britain had the competence to provide proper mental and moral training, moved him to take charge of his son's tuition. James Mill, schooled in the psychology of Locke and Hartley, held that all mental phenomena sprang from sensory elements and could be explained by the principle of association. The simultaneous or successive experience of sensations produces simultaneous or successive ideas. The process gives rise to clusters of ideas and we cease to be aware of the simple elements from which the clusters are constructed. The primary duty of the educator was to foster in his pupil the formation of salutary associations. James Mill had an unbounded confidence in the capacity of education to promote moral and mental excellence, and the education of his son was in part intended to demonstrate the validity of this notion.[11]

Less than two months after the birth of this son James Mill wrote a letter to William Forbes, who had married the daughter of Sir John Stuart. Like Mill himself, Forbes had recently become a father. The elder Mill's pleasure in his own changed status is palpable—"The boy is as fine a child as possible"—as is his desire to show what he could make of the infant responsible for that pleasure.

> I intend to run a fair race with you in the education of a son. Let us have a well-disputed trial which of us twenty years hence can exhibit the most accomplished & virtuous young man. If I can beat you in this contest, I shall not envy that you have yours the richest. I know not how far I may fall from my good resolves in this as I do in most other cases, but I have a strong determination at present to exert myself to the utmost to see what the power of education can do.[12]

James Mill's earnest and unfaltering perseverance in this undertaking is classically attested to in his son's *Autobiography*. Nearly thirty-five years before that document began to take shape, the elder Mill summarized for a correspondent the course of John's education from infancy to adolescence.

> Being convinced of the advantages which a father enjoyed in swaying the mind of his child, & being occupied wholly at home when I first became a father, I began with my first child, a son. The principle of imitation, seeing books my grand occupation, made his curiosity attach itself to books; & when in our little intercourse he

desired to look at a book, instead of showing him pictures, I showed him letters. In this way, without any trouble he knew the letters, & more (how much more I do not recollect) before he was 18 months old—& before he was three years old he could read English perfectly. The same principle of imitation led his curiosity to Greek books, owing to the novelty of the character. I availed myself in like manner of this curiosity to make him acquainted with the Greek characters, & after that with the inflections of the nouns & verbs. In the mean time he was occupied with maps; & by the time he was five years old, he knew a good deal of Greek, & was acquainted with geography even to minuteness. Greek went on, & reading simple books of history, for perhaps a couple of years, during which he had begun arithmetic. With Latin . . . he became acquainted very rapidly, having first learned Greek. In this way, he has gone on; & from no part of his time having been allowed to go to waste, his acquirements are very unusual at his years. He is not 14 years old till next may [sic]—& he is not only a good Greek & Latin scholar, but he has actually read, almost all the Greek & Latin classics—he is well versed in Mathematics even fluxions & the higher branches—I know of nobody who has in his memory a greater amount of historical facts—I have taught him Logic, & I have taught him political economy . . . [H]e has also a good knowledge of chemistry. His studies were always carried on in company with me. He sat in my room, & studied when I studied—& though attending to him when he needed it produced some interruption, yet all I have done, was done under it.[13]

James Mill was satisfied that "no part of" his son's time had "been allowed to go to waste"; in his *Autobiography* the son remarks upon the "wretched waste of so many precious years as are spent acquiring the modicum of Latin and Greek commonly taught to schoolboys," and credits his father with having taken "great effort to give, during the years of childhood, an amount of knowledge in what are considered the higher branches of education, which is seldom acquired (if acquired at all) until the age of manhood."[14] This education, moreover, "was not an education of cram." James Mill, his son notes, "strove to make the understanding not only go along with every step of the teaching, but if possible, precede it. Anything which could be found out by thinking, I never was told, until I had exhausted my efforts to find it out for myself." Exhausted he must often have been. James Mill, it seems, did not hesitate to show where his son's understanding had proven deficient.

The pupil's "recollection of such matters is almost wholly of failures, hardly ever of success."[15] The learning environment created by James Mill wore a menacing aspect. Francis Place, who greatly admired James Mill, had an opportunity to observe the latter's instructional practices in August 1817. "His method is by far the best I ever witnessed, and is infinitely precise; but he is excessively severe. No fault, however trivial, escapes his notice; none goes without reprehension or punishment of some sort."[16] Discussion of facts and ideas certainly occurred; one of the parties, however, dominated the discussion, through force of character and threatening demeanor as much as through cumulative experience and power of intellect.

Soon after the education of John Mill began, James Mill became acquainted with Jeremy Bentham. In 1808 James Mill wrote an article for the *Annual Register* that bestowed high praise on Bentham, whose work was "entitled to profound regard. Of all the men, in all ages, and in all countries, who have made philosophy of law their study, he has made the greatest progress."[17] To be admired by a man of Mill's faculties appealed to Bentham's vanity. Although Mill was a man with projects of his own who deeply valued his personal independence, he considered the propagation of Bentham's doctrines a matter of supreme importance. He would become Bentham's most influential disciple. From 1809 to 1818 their friendship was a central element in the lives of both men. In 1810 Mill agreed to live in a Westminster house that belonged to Bentham (this house, close to Bentham's garden, had once been occupied by Milton). When this home proved unsatisfactory, Mill moved his family to Newington Green, a change that did not prevent him from visiting Bentham frequently at the latter's house in Queen Square Place. Such visiting was made a good deal easier in 1814 when Bentham persuaded Mill to become his neighbor by offering to pay half the rent on a house at 1 Queen Square. (In October 1813 the government had made Jeremy Bentham a rich man by awarding him £23,000 in compensation for failing to implement his Panopticon prison scheme.) From 1814 through 1818 Mill and his family also spent about half of each year with Bentham at Ford Abbey, a large Somersetshire country house near Chard.

The association was not without strain, the two men's personalities and habits being strikingly dissimilar. Austere, efficient, and resolute, Mill applied his sustained concentration to what he took to be his duties. Bentham's waywardness formed a marked contrast. Although the great man sought to provide comprehensive and systematic coverage of the many topics he addressed, he completed but a small portion of his ambitious undertakings, signaling a personal constitution subject to fits of caprice and distraction. So, at least, it often must have seemed to James Mill. Yet he remained unswervingly loyal to the Benthamite cause of fundamental legal and political reform. His rendering of that cause, in discussion and in print, gave it a tone and cogency that a band of young acolytes would find compelling. (In Elie Halévy's words, "Bentham gave Mill a doctrine, and Mill gave Bentham a school.")[18] James Mill's eldest son would, for a time, be among the most strident of these adherents. The father expected no less. The training so assiduously given his first child was designed, in large part, to cultivate a mind capable of carrying forward the Benthamite program after Jeremy Bentham and James Mill were gone.[19]

A set of sharply defined suppositions concerning human nature and society underpinned this program. Bentham assumed that all actions arising from the exercise of human will stemmed from a desire to procure some pleasure or escape some pain (he did not fail to see that the range of pleasures and pains motivating individuals was considerable).[20] Although he recognized that sympathy and benevolence—non-self-regarding pleasures—could affect human conduct, he nonetheless held that "in the general tenor of life, in every human breast, self-regarding interest is predominant over all other interests put together."[21] This Bentham did not lament, for the survival of the human species had depended upon the operation of the "self-preference" principle. Human nature itself prescribed that each individual should seek to maximize his own happiness or pleasure, and Bentham's system made no room for the notion that human nature either could or should be altered. The defining characteristic distinguishing his concept of utility from that of Hume or Helvetius was "the greatest happiness principle." The object of a healthy social order should be "the greatest happiness of the greatest number," and the value of all actions should be measured in relation to this end. Bentham, however, did not suppose that individuals would in fact attempt to make such calculations. The pursuit of individual self-interest, to be encouraged as a rule, could conflict with the larger interest of the community in promoting the greatest happiness principle. Government had been instituted to protect life and property; its aim was to give individuals the security needed to advance their self-interests without fear of violence or arbitrary expropriation. Application of the greatest happiness principle was chiefly to be entrusted to legislators, who had the responsibility to put in place the legal and institutional arrangements favorable to increasing the aggregate quantity of happiness.[22]

This task involved devising the means that would induce individuals to seek happiness in ways that would not harm others; better yet, these means should encourage conduct contributive to the happiness of others. Inasmuch as people were largely driven by desire for pleasure and aversion to pain, rewards and punishments—or the prospect of such—were the instruments to be applied by the legislator. The main role of rewards was to stimulate individuals to engage in activity beneficial to the public. The activity at issue and the rewards available were limited in nature and could not be relied upon to elicit a response from the great bulk of the population. Clubs outnumbered hearts in the deck from which the legislator had to draw. Bentham wished to give people motives for refraining from self-regarding actions whose consequences would be socially injurious: actions that would reduce the community's sum of happiness. Only the threat and, where unavoidable, imposition of penalties could furnish such motives.[23]

The realization of Bentham's schemes for radical legal reform therefore hinged on the presence of a legislature with the competence and public spirit to see what needed to be done and to act accordingly. As of 1809, Bentham had become convinced that the British Parliament, as then constituted, was far too corrupt to perform this function (the frustration occasioned by the government's handling of his Panopticon project probably had some bearing on this judgment). The landed classes controlled the political system, and they unsurprisingly used their predominance to advance their own interests at the expense of the rest of the community (so Bentham had concluded). The legislature would serve the general interest—the aggregate of all individual interests—only when the representative system allowed for the effective expression of that interest. Bentham thus embraced political radicalism. In his *Plan of Parliamentary Reform* (1817) he advocated universal suffrage, annual parliaments, and the secret ballot.[24]

It is almost certainly no accident that Bentham's turn toward political radicalism coincided with the formation of his close association with James Mill. The latter, believing ignorance to be the chief obstacle to progress, devoted much thought and energy to educational reform. The masses lacked the knowledge to understand that their long-term happiness called for them to discipline their bodily appetites, which they were inclined to gratify immediately and to excess whenever opportunity presented. Their rulers wanted to keep them in this condition; were the people to know the real sources of their misery they would insist on the institutional reforms necessary for a proper calculation to be made of the public good. Such reforms would inevitably deprive the aristocracy of their political ascendancy and roll back their pernicious influence. Much of James Mill's political writing was directed at showing the better educated among the middling elements that the perpetuation of aristocratic power was incompatible with social progress. He presumably had a part in persuading Bentham that such was the case.

These ideas turned up regularly in the discussions James Mill had with his eldest son. J.S. Mill notes that his education had, "in a certain sense," been "a course in Benthamism," this sense being that he had "always been taught to apply" Bentham's greatest happiness principle.[25] Yet his response in the winter of 1821–22 to reading the *Traité de Législation*, Etienne Dumont's redaction of Bentham's *Introduction to the Principles of Morals and Legislation*, implies that his education had been less doctrinaire than might be thought. In his *Autobiography* he writes: "The reading of this book was an epoch in my life; one of the turning points in my mental history." He amplifies: "when I laid down the last volume of the *Traité* I had become a different being. The 'principle of utility,' understood as Bentham understood it, and applied in the manner in which he applied it . . . fell exactly into its place as the keystone which held together the detached and fragmentary parts of my knowledge and beliefs."[26] Had his education been rigidly Benthamite in character, his reading of Dumont's redaction would not have had such a transformative effect.

Although his learning and convictions before 1821–22 were "detached" and "fragmentary," the knowledge and understanding he acquired as a youth and brought to his reading of Bentham possessed a startling range and depth for one of his age. The beneficiary of this education, ingenuously or otherwise, rejected the notion that he possessed exceptional abilities.[27] Yet the tutelage of James Mill and the ample library of Jeremy Bentham clearly found fertile soil in John Mill's powerful young mind. Even James Mill had to acknowledge this power, if only by implication. Just a couple of years before reading the *Traité*, John had received from his father "a complete course of political economy."[28] The outcome proved satisfactory to the preceptor himself, whose pride in both the instruction on offer and the intellectual equipment of the pupil is obvious. "I have taught him Logic, & I have taught him political economy. Mr. Ricardo [Ricardo's seminal *On the Principles of Political Economy and Taxation* had been published in 1817] who was interrogating him the other day, says he knows nobody by whom even the most abstruse points of the science are better understood."[29] At the time this was written his son was still short of his fourteenth birthday. John Mill's education gave him, by his own account, "an advantage of a quarter of a century over his cotemporaries."[30] James Mill's aim was to give his eldest son the means to achieve intellectual and moral excellence. The training he imposed was not intended to school his pupil in a "system" of thought that was to be accepted uncritically. The unfettered and impartial exercise of the critical intelligence could not yield false conclusions. In his *Liberty of the Press* (1821) James Mill declared: "Every man, possessed of reason, is accustomed to weigh evidence, and to be guided and determined by its preponderance."[31] If James Mill had his way, no one would weigh evidence with greater acuity, discrimination, and understanding than his eldest son, who would arrive at the conclusions arrived at by his father simply because they were true.

The intellectual excellence James Mill fostered in John would enable him to learn what was true; the moral excellence he sought to instill would make John's actions consistent with his knowledge. In James Mill's 1806 letter to William Forbes, quoted above, reference is made to a contest between the two men to see who could produce "the most accomplished & virtuous young man." A letter written some fifteen years later alludes to James Mill's hopes for his son's "making a shinin[g] character."[32] For Bentham and James Mill the greatest happiness principle was the standard by which actions should be judged. In some respects, however, James Mill was an odd sort of utilitarian. "[H]e had scarcely any belief in pleasure," his son reports. James Mill ascribed the "greatest number of miscarriages in life . . . to the overvaluing of pleasures."[33] By the time the education of John had begun James Mill had ceased to be a Christian (if he had ever truly been one) and had become a skeptic; the puritan strain characteristic of the Presbyterianism in which he had been reared nonetheless retained a conspicuous hold on him. "My father's moral inculcations," John states, "were at all times mainly those of the 'Socratici viri'; justice, temperance (to which he gave a very extended application), veracity, perseverance, readiness to encounter pain and especially labour." James Mill esteemed "a life of exertion, in contradiction to one of self-indulgent sloth."[34] As his invoking of the "Socratici viri" suggests, John Mill was disposed to link these attributes to the influence of the ancients on his father's convictions and character. The linkage may well be justified, but the influence worked upon a sensibility that bore the imprint of a Presbyterian upbringing.[35]

The moral instruction James Mill conveyed to his son was not free of ambiguity. The Socratic virtues valued by James Mill would also be prized by John Mill. Together they read Xenophon's *Memorabilia*, upon which James Mill offered comments. From this experience the younger Mill came to see Socrates "as a model of moral excellence."[36] John Mill recalled how his father "impressed upon me the lesson of the 'Choice of Hercules,'" a choice between *happiness*, as commonly (or vulgarly) understood, and *virtue*. James Mill made plain his strong preference for the latter. Free and fearless inquiry also ranked high in James Mill's estimation. Although John Mill was "brought up from the first without any religious belief,"[37] his father taught him "to take the strongest interest in the Reformation, as the great and decisive contest against priestly tyranny and liberty of thought."[38] All the same, James Mill cautioned his son against revealing his want of religious faith. Better to leave unsaid certain things one thought. In early-nineteenth-century England, blasphemy was a crime (the common-law offense still exists); the real threat of prosecution and incarceration aside, declarations of hostility to Christianity could mark a person as irresponsible and unsafe. (Arguing vigorously for freedom of religious expression was a different matter.) The young John Mill, it seems, had trouble adhering to his father's injunction, which had more the flavor of temporizing than of temperance (John, it must be said, would not have put it this way). Had not one of the charges against the heroic Socrates been blasphemy? No one in James Mill's England bold enough to take on Christianity had to worry about suffering the punishment imposed on Socrates. Refusal to avow what one believed for fear of censure smacked of hypocrisy and cowardice, and John Mill remembered two instances from his youth when he felt obliged to reject his father's advice. "My opponents were boys, considerably older than myself: one of them I certainly staggered at the time, but the subject was never renewed between us: the other, who was surprised and somewhat shocked, did his best to convince me for some time, without effect."[39]

The mature John Mill tended to heed his father's advice, notwithstanding his assertion in the *Autobiography* that there had been a "great advance in liberty of discussion" since the early decades of the nineteenth century. Men of James Mill's strong convictions, intelligence, and devotion to the public good should now be forthright in stating their religious views, however unpopular those views might be. Yet, even in the more enlightened mid-nineteenth century, reticence was justified if an individual risked losing his livelihood or faced "exclusion from some sphere of usefulness peculiarly suitable to the capacities of the individual."[40] The second exception had to be allowed for J.S. Mill to satisfy his own scruples. As late as the mid-1860s he refused to say publicly that he was not a Christian. When facing the electors of Westminster at the 1865 general election, he pointedly declined to answer any question bearing upon his religious creed.[41] In November

1868, in the midst of allegations of atheism spawned by his donation to the campaign of England's most notorious atheist, Charles Bradlaugh, Mill went so far as to send a letter to the newspapers in which he defied anyone to cite a single passage from his voluminous writings in support of such a charge.[42] Mill's most direct remarks on his religious position appear in the *Autobiography* and in his *Three Essays on Religion*, both published posthumously.[43] The moral training given John Mill had enduring effects. The conduct it inculcated was not all of a piece.

★ ★ ★

A pause in James Mill's direct supervision and surveillance of his son's education occurred in 1820. In May John left England for France, where he spent much of the next year. He had been invited to stay with Sir Samuel Bentham and his family, who had taken up domicile in Pompignan, not far from Toulouse. Sir Samuel, the younger brother of Jeremy, was a remarkable individual in his own right. A brilliant and versatile man, he had achieved some distinction as a naval architect, inventor, and mineralogist. He had done a stint as a brigadier general in the army of Russia's Catherine the Great, and another as Inspector General of His Majesty's Naval Works. His wife, Maria Sophia Bentham, was the daughter of George Fordyce, an eminent Scottish physician and chemist. A highly cultivated and confident woman, Lady Bentham managed the household and organized and superintended many of John Mill's activities during his stay with the family. This lively household included three daughters and a son. The latter, George, John's senior by six years, had displayed an intellectual precocity little less spectacular than that of his younger guest. He became one of the nineteenth century's foremost botanists (John Mill's abiding enthusiasm for botany began during his sojourn with the Benthams). Here was a family that could appreciate John Mill's uncommon attainments without being awed by them.[44]

John was evidently unaware that any family should be awed by these attainments. He notes that he "had no notion of any superiority" in himself. James Mill had done his utmost to ensure that his eldest son would not develop a lofty view of his own accomplishments. The sermon he gave John shortly before the latter's departure for France was characteristic of the elder Mill's efforts in this sphere. John was informed that he could claim no credit for knowing more than other boys his age; that he did so sprang from his "very unusual advantage . . . in having a father who was able to teach" him and who was "willing to give the necessary trouble and time." The idea that he might deserve "praise" for what he had learned cut no ice with James Mill. In light of John's privileged educational circumstances, it would have been "the deepest disgrace" to him had he failed to acquire more knowledge than his contemporaries. Owing to these circumstances, it should be said, John had been kept ignorant of his contemporaries' relative ignorance. He assures readers of his *Autobiography* that he never felt disposed to congratulate himself (self-congratulation would have been a remarkable feat for a recipient of James Mill's lavish attention); his father's admonition concerning the praiseworthiness of his achievements met with a ready assent. Now that the subject of his "acquirements" had been broached by James Mill, John "felt that what my father had said respecting my peculiar advantages was exactly the truth and common sense of the matter, and it fixed my opinion and feeling from that time forward."[45]

Not long after this conversation John caught the Dover coach and headed for France. Some twenty years later he would refer to his months abroad in 1820–21 as "les plus heureux de ma jeunesse."[46] It is easy to see why. He had the company of the convivial and stimulating Bentham family; opportunities to view sublime scenery of a kind altogether different from what he had previously encountered; enjoyable botanical expeditions with George Bentham; meetings with French radicals and intellectuals; the satisfaction of winning a growing mastery of the French language; the bracing experience of attending courses at the University of Montpellier; and the nourishment—so it was to the eldest son of James Mill—of an ambitious program of study.[47] Contributing to this happiness, there is reason to think, was the absence of James Mill. Not that his father could have been absent from the boy's thoughts. James Mill had told his son to keep a record of all that he did, and John understood for whom the record was being kept. Simply to be out from under his father's gaze, however, must have lessened considerably the anxiety aroused by his hitherto pervasive presence.

A passage that Harriet Taylor Mill cancelled from the Early Draft of her husband's *Autobiography* offers a glimpse of the shell the son constructed in response to this gaze.

> I . . . grew up in the absence of love and in the presence of fear: and many and indelible are the effects of this bringing-up, in the stunting of my moral growth. One of these, which it would have required a quick sensibility and impulsiveness of natural temperament to counteract, was habitual reserve. Without knowing or believing that I was reserved, I grew up with an instinct of closeness. I had no one to

whom I desired to express everything which I felt; and the only person I was in communication with, to whom I looked up, I had too much fear of, to make the communication to him of any act or feeling ever a matter of frank impulse or spontaneous inclination.[48]

This may, or may not, be understatement; assuredly it is not exaggeration. John Mill was a dutiful, loyal, and grateful son. Filial piety informs most of what he wrote about his father. In a January 1854 journal entry, composed about the same time the Early Draft was under way, J.S. Mill protested at the failure of his own generation to appreciate his father's true stature: "Who was ever better entitled to take his place among the great names of England? He worked all his life long with complete disinterestedness for the public good; he had no little influence on opinion while he lived, most of the reforms which are so much boasted of may be traced mainly to him, and in vigour of intellect and character he stood quite alone among the men of his generation."[49] His veneration for his father's memory did not keep John Mill from remembering that he had grown up "in the absence of love and in the presence of fear."

There is no open expression of resentment; no urge to assign blame. John Mill would not believe that his father's want of tenderness "lay in his own nature." Like Englishmen generally, James Mill repressed his feelings from a sense that to display them was unmanly. Habitual repression led to emotional deprivation. When allowance was made for his being in the "trying position of sole teacher," and for his "constitutionally irritable" temperament, who could not "feel true pity for a father who did, and strove to do, so much for his children, who would have so valued their affection, yet who must have been constantly feeling that fear of him was drying it up at its source"?[50]

James Mill's unwavering devotion to duty in the face of vexing circumstances may well have entitled him to such retrospective consideration. He chose to be his eldest son's "sole teacher"; in teaching most of the children who followed John, James Mill had an able assistant. Yet his position was "trying" enough. For a number of years he undertook the task of instructing John while struggling to support a rapidly growing family, enduring a marriage that brought him no joy and some exasperation, and doggedly carrying on with his *History of British India*. In showing imaginative sympathy for his father's predicament, the mature J.S. Mill also shows a sensibility superior to that possessed by the object of his pity. James Mill, one supposes, would have bristled at his son's patronizing tone. The notion that he was conscious of the fear he evoked in his children is perhaps the most striking aspect of the passage quoted above. How someone whose feelings had been enfeebled could have "constantly" felt that the fear he kindled had deprived him of his children's affection is hard to fathom. The treatment of this fraught subject in the pages of the *Autobiography* reveals much about the difficulty J.S. Mill had in coming to grips with the memory of a father who had been both the great intellectual benefactor and the oppressive emotional bane of his childhood. Compassion for James Mill's plight could be more easily mustered after his death.

The weight of James Mill's authoritarian demeanor must have put considerable stress on his son's psychological constitution. A robust example of James Mill's peremptory manner is found in a letter he wrote to his great friend David Ricardo in 1818. By this time Ricardo, James Mill's senior by one year, had established himself as a brilliant and influential political economist. The subject of the communication was the method Ricardo should adopt to master the composition of an effective argument. James Mill told his friend that the "discourses" they had spoken of should "be written without delay." He then gave a detailed and sequential outline of the mode by which Ricardo must proceed. The imperative voice richly informed James Mill's closing remarks.

> One thing more, however; you must write your discourses, with the purpose of sending them to me. Depend upon it, this will be stimulus, not without its use. I will be the representative of an audience, of a public; and even if you had in your eye a person whom you respect much less than you do me, it would be a motive both to bestow the labour more regularly, as it should be; and to increase the force of your attention. Therefore no apologies, and no excuses will be listened to.[51]

Ricardo did not live with James Mill. The young John Mill could not opt to have anyone other than his father in his "eye." And what he saw could not fail to command respect and incite fear. James Mill gave his son a compelling "motive" to engage in rigorous study; his vigilant superintendence of John's education powerfully concentrated the "force" of his son's "attention." The dictatorial implementation of this program cost James Mill his son's affection. John Mill too paid dearly.

In the *Autobiography* J.S. Mill says that his father's "severity . . . was not such as to prevent me from having a happy childhood."[52] That this "severity" detracted from his happiness is plain. James Mill's harshness caused some suffering, and it is arguable that this harshness was unnecessary to the achievement of the ends he had in view. John Mill does not suggest this. Indeed, he asserts that "Much must be done, and much must be learnt, by children, for which rigid discipline, and known liability to punishment, are indispensable as means."[53] This conviction might have issued from the frustration John Mill experienced in teaching his younger siblings. It was also born of the need to believe that a large part of what his father had put him through had been essential to his intellectual development. Yet it is reasonable to think that John would have accomplished a great deal under a much milder regime. His father undoubtedly stimulated and fed his hunger for knowledge. The hunger itself, however, belonged to the son. His love of learning, coupled with his exceptional ability, made him an unusually apt pupil. The tractability he demonstrated did not stem from a "rigid discipline" imposed from without or from a "known liability to punishment." He wanted to learn as much as he could, and thus he was "happy" despite the exacting and imperious rule of his father. Such a reflection renders all the more poignant the passage in the *Autobiography* where Mill reckoned the cost of what he had endured.

> I do not, then, believe that fear, as an element in education, can be dispensed with; but I am sure that it ought not to be the main element; and when it predominates so much as to preclude love and confidence on the part of the child to those who should be the unreservedly trusted advisers of after years, and perhaps to seal up the fountains of frank and spontaneous communicativeness in the child's nature, it is an evil for which a large abatement must be made from the benefits, moral and intellectual, which may flow from any other part of the education.[54]

Fear did so predominate in the case of John Mill.

★ ★ ★

In his autobiographical account J.S. Mill notes that when he returned from France his father no longer acted as his "schoolmaster." John's education nonetheless remained, for a time, under James Mill's "general direction."[55] During the summer and autumn of 1821 James Mill was readying for publication his *Elements of Political Economy*, a book that drew heavily on summary notes written by John in 1819 in connection with a course of instruction he had received from his then "schoolmaster." His father had "expounded each day a portion of the subject, and I gave him next day a written account of it, which he *made me* [italics added] rewrite over and over again until it was clear, precise, and tolerably complete."[56] Apparently the change in regimen coinciding with John's return home did not include a softening in James Mill's tone. The younger Mill observes that his father then "made" him "perform an exercise on the manuscript" that involved composing a brief abstract of each paragraph in the margin. Although James Mill doubtless thought this would be a valuable "exercise" for the performer, its primary aim, we are led to infer from his son's account, was to serve the author of the manuscript, who could thereby "more easily . . . judge of, and improve, the order of the ideas, and the general character of the exposition."[57]

James Mill's penchant for judging, improving, and ordering meant that he would decide the choosing of a livelihood for his eldest son. John's "calling," that of Benthamite reformer, could not be expected to provide for his wants. The practice of law, his father surmised in 1821, was perhaps the least objectionable answer to the problem. Bentham, the most comprehensive and acute critic of English law, had himself been called to the bar (not that he could be said to have established much of a practice thereafter). Arrangements were made for John to read Roman law with John Austin during the winter of 1821–22. John and Sarah Austin had moved from Norwich to London following their marriage in 1819, becoming neighbors of Bentham and the Mills in Queen Square Place. An ardent admirer of Bentham's legal doctrines, Austin at this time was also a keen supporter of Bentham's radical political program (he would later disavow democratic radicalism). His mind had power and originality—J.S. Mill remarked that Austin "had made Bentham's best ideas his own, and added much to them from other sources and from his own mind"—and his talk often exhibited an astringent fluency.[58] His somber and volatile temperament found expression in both a dogmatic and agitated assertiveness and in episodes of incapacitating despair and lethargy. Austin's intellect and character made a deep impression on John Mill. "On me his influence was most salutary. It was moral in the best sense. He took a sincere and kind interest in me, far beyond what could have been expected towards a mere youth from a man of his age, standing, and what seemed austerity of character."[59] The strong intellect and amiable personality of Sarah Austin also had an effect on John Mill. Attractive, clever, engaging, and sympathetic, she warmed his existence in ways he would not acknowledge when he came to write his autobiography.[60]

The Austin household offered consolation to John Mill for having to leave the home of Sir Samuel Bentham and his family and return to that of his father. John Austin accorded the youth a pleasing respect, while

Sarah Austin showed him great cordiality and affection. John Mill became a favorite with their delightful infant daughter Lucy, to whom he was known as Bun Don (her version of Brother John). He joined the Austins for holidays, which, among other advantages, placed him temporarily beyond the reach of his father. The house occupied by James Mill held for his eldest son troubling associations: a feared parental presence; a mother he could not respect; siblings he had the duty to instruct. Fortunately for John, James Mill approved of the Austins, who were seen as aids rather than obstacles to his son's intellectual and moral growth (he entrusted his daughters' lessons to Sarah Austin while John was in France). For John Mill the Austin household was a source of edification, enjoyment, comfort, and encouragement.

Reading law with John Austin had a lasting effect on John Mill's intellectual development; getting to know Charles Austin, John's younger brother, had an immediate impact on John Mill's personal development. In the *Autobiography* Mill says that "at this time and for the next year or two" he "saw much" of Charles Austin. The initial meeting took place in the summer of 1822, when John Mill was on holiday with the Austins in Norwich. John Mill's senior by seven years, Charles Austin had made a brilliant student career at Cambridge, where his oratorical and conversational power dazzled his contemporaries (Thomas Macaulay included).[61] His spellbinding performances at the Cambridge Union gave wide exposure to Benthamite doctrines within the university. Thirty years later J.S. Mill wrote that Charles Austin's influence at Cambridge "deserves to be accounted an historical event; for to it may be traced the tendency towards Liberalism in general, and the Benthamic and politico-economic form of it in particular, which shewed itself in a portion of the more active-minded young men of the higher classes from this time to 1830."[62] The extraordinary potency of Charles Austin's personality resonates vividly in an account far removed in time from the events that had introduced him to the young John Mill.

> The impression he gave was that of boundless strength, together with talents which, combined with such apparent force of will and character, seemed capable of dominating the world . . . He loved to strike, and even to startle. He knew that decision is the greatest element of effect, and he uttered his opinions with all the decision he could throw into them, never so well pleased as when he astonished any one by their audacity. Very unlike his brother, who made war against the narrower interpretations and applications of the principles they both professed, he on the contrary presented the

Benthamic doctrines in the most startling form of which they were susceptible, exaggerating every thing in them which tended to consequences offensive to any one's preconceived feelings. All which, he defended with such verve and vivacity, and carried off by a manner so agreeable as well as forcible, that he always either came off victor, or divided the honours of the field.[63]

Had Charles Austin penned a portrait of the adolescent John Mill, the details would surely have differed from those in the passage above; the trenchancy of the impression made might well have been comparable. Eager to show off this remarkable prodigy to his Cambridge friends, Charles Austin arranged for John to meet T.B. Macaulay, the brothers Hyde and Charles Villiers, Edward Strutt, and John Romilly. All were active in the Cambridge Union, and all, except for Hyde Villiers, who died young, were to play notable parts in Victorian public affairs.[64] This marked the first time John Mill was thrown together with a group of worldly and able young men against whom he could measure his own attainments. He found that he could more than hold his own. The regard Charles Austin showed John Mill clearly had a significant effect on the latter. Mill's *Autobiography* notes that Charles Austin's "influence . . . over me differed from that of the other persons I have hitherto mentioned, in being not the influence of a man over a boy, but that of an elder cotemporary. It was through him that I first felt myself, not a pupil under teachers, but a man among men. He was the first person of intellect whom I met on a ground of equality."[65] John Mill had never been a boy among boys; the first experience of feeling himself "a man among men" was unforgettable. His feeling this way, at the age of sixteen, and in such a milieu, shows that his unusual upbringing had not rendered him unfit for satisfying social and intellectual intercourse with those primed to assume places of prominence in the public life of nineteenth-century England. Reconciling life as "a man among men" with life as the eldest son of James Mill would call for the expenditure of much emotional energy over the course of the coming years.

For James Mill, John's legal studies with John Austin served a dual purpose: preparing his son for a possible career at the bar while advancing his education in Benthamism. At the outset of John's course of instruction with Austin, James Mill placed in his hands the *Traité de Législation*, "as a needful accompaniment."[66] The reading of this work gave John Mill "a creed, a doctrine, a philosophy; in one among the best senses of the word, a religion; the inculcation and diffusion of which could be made the principal outward purpose of a life." His father is given credit for having laid the foundation for his son's reception of the *Traité*, but it took Bentham's treatise to bring "unity" to his "conceptions of things."[67] It was the gospel according to Bentham that became, for a time, his lodestar. The partial transfer of authority from James Mill to Bentham perhaps intensified the exultation John Mill experienced upon reading the *Traité*.

Another aspect of John Mill's philosophical inheritance assumed a more definite shape soon after his encounter with the *Traité*. In 1822 his father closely supervised his reading of John Locke's *Essay Concerning Human Understanding* and David Hartley's *Observations on Man, His Frame, His Duty and His Expectations*. During the summer of that year James Mill began writing what would become his *Analysis of the Phenomena of the Human Mind*, which "carried Hartley's mode of explaining the mental phenomena to so much greater length and depth."[68] John Mill's systematic study of "analytic psychology" fixed in his mind the irrefutable authority of the law of association. The experiential school with which he identified rejected the idea that there existed truths outside the mind that could be apprehended by intuition. All knowledge issued from experience. Ideas regarding external objects were formed from "sensations received together so frequently that they coalesce . . . and are spoken of under the idea of unity."[69] The "sensations received" by intuitionists had regrettably duped them into supposing that their ideas corresponded to an innate reality. Many of the sensations reaching the young John Mill had been sent by his father. Associationism's deterministic streak did not, it seems, cause John problems in 1822, when it can be said his foundational theories— Benthamite utilitarianism in ethics and associationism in psychology— were set in place. For several years thereafter a surge of practical activity absorbed a large portion of his energies and helped ward off the melancholy reflections that would attend his mental crisis.

His reliance on his father and Bentham in the realm of theory did not stop John Mill from being a leader of sorts in these years. In late 1822 and early 1823 he started up the Utilitarian Society, which consisted of a small number of young Benthamites who met twice a month at Bentham's house "to read essays and discuss questions."[70] His coadjutors in this endeavor, Mill notes, were "less advanced" than himself, and he "had considerable influence on their mental progress."[71] In 1824 he was instrumental in the formation of the Society of Students of Mental Philosophy, which focused on philosophical and economic questions. The creation of the London Debating Society in 1825 also owed something to John Mill's energy and commitment. He played an important role in the planning and launching of the *Parliamentary History and Review*, a short-lived periodical of the mid-1820s that propagated a sectarian Benthamite line on topics taken up by Parliament. Apart from these activities, he was also contributing pieces to the newspapers, writing frequent articles for the *Westminster Review* (established by Bentham in 1823–24), and editing Bentham's *Rationale of Judicial Evidence*, which appeared in five volumes in 1827. Whatever he might subsequently say about the deficiencies of his education, the young John Mill plainly had a notable capacity for acting on his beliefs.[72]

In all this there is nothing to suggest that the adolescent John Mill would have shared the view of the mature J.S. Mill that his "own strength lay wholly in the . . . region . . . of theory."[73] The drama and romance of life resided in the sphere of action, a fact the author of the *Autobiography* makes no attempt to hide when writing of youthful yearnings. These focused on the performance of great acts meriting public acclaim and gratitude. In the early 1820s, not long after his return from France, John Mill developed a keen interest in the history of the French Revolution. Having had but a sketchy knowledge of the subject before this time, he "learnt with astonishment, that the principles of democracy, then apparently in so insignificant and hopeless a minority everywhere in Europe, had borne all before them in France thirty years earlier, and had been the creed of the nation." This discovery carried a distinct emotional impact: "the subject took an immense hold of my feelings. It allied itself with all my juvenile aspirations to the character of a democratic champion. What happened so lately, seemed as if it might easily happen again; and the most transcendant [sic] glory I was capable of conceiving, was that of figuring, successful or unsuccessful, as a Girondist in an English Convention."[74]

The exemplary status assigned the Girondists by the youthful John Mill had no place in the lessons he learned from his father. The post-1792 reaction in England against the French Revolution was of such intensity that a defense of the ineffectual political rivals of the Jacobins had nothing to recommend it. For most of the decade following John Mill's birth, the United Kingdom was at war with an immensely powerful Napoleonic France. Bonaparte's rise, as understood in England, had been made possible by a revolution destined to bring destruction in its wake. Albeit lacking the zealotry of the Jacobins, the Girondists nonetheless were easily blackened with the republican brush. A person seeking to acquire a measure of influence in England during the early decades of the nineteenth century would be well-advised to say nothing in favor of any kind of revolutionary. In his early years in London James

Mill himself had written articles for the *Anti-Jacobin Review*. His subsequent political radicalism had no use for the *Declaration of the Rights of Man*, whose language, from a Benthamite perspective, seemed riddled with fallacies. With the restoration of the Bourbons, James Mill entertained the idea of moving his large family to France owing to its relative cheapness. To his friend Francis Place he wrote: "I foresee nothing there which would make it uncomfortable for us to reside as soon as we please. Assure yourself that the French people will soon be very quiet & contented slaves, & the despotism of the Bourbons a quiet, gentle despotism."[75] After his stay with the family of Samuel Bentham, and his exposure to the history of the French Revolution, the one thing that could make France an unendurable place for John Mill to reside would be a conviction that the French people had become "quiet & contented slaves."[76]

The emphasis given analysis in his early education had clearly not crippled John Mill's imaginative faculties. The renown he wished to win from acting the part of a Girondist in an "English Convention" signals an appetite for heroic conduct and an eagerness to visualize the circumstances summoning this heroism. Although a mature J.S. Mill situated this longing within a congeries of "juvenile aspirations," more than a trace of such enthusiasms can be detected decades after he had left childhood behind. For all James Mill's detestation of aristocratic government, for all his investment in training his eldest son to become a benefactor of mankind, the idea of John Mill's participation as a Girondist in an "English Convention" is one he did not reckon with.

Reading the history of the French Revolution fired the younger Mill's imagination and aroused a powerful emotional response in one who aspired to distinguish himself in the sphere of action no less than that of thought. Nothing in this history, regrettably, could illustrate the effective coupling of theory and practice, and James Mill had no reason to direct his son toward the French Revolution with a view to supplying him with a positive object lesson. For an exemplary illustration of the integration of thought and action, both Mills looked not to eighteenth-century France but to ancient Greece. In his 1832 essay "On Genius," John Mill purports to describe this special historical moment.

> The studies of the closet were combined with, and were intended as a preparation for, the pursuits of active life. There was no *litterature des salons*, no dilettantism in ancient Greece: wisdom was not something to be prattled about, but something to be done. It was this which, during the bright days of Greece, prevented theory

from degenerating into vain and idle refinements, and produced that rare combination which distinguishes the great minds of that glorious people,—of profound speculation, and business-like matter-of-fact common sense . . . Bred to action, passing their lives in the midst of it, all the speculations of the Greeks were for the sake of action, all their conceptions of excellence had a direct reference to it.[77]

This passage may reveal less about the ancient Greeks that it does about John Mill's education and sensibility. In the early 1820s, unlike the early 1830s, he could not be said to have "speculations" that were entirely his own. His imagination, even then, he did own, and his Girondist fantasy shows he wanted what he had learned to be "for the sake of action."

★ ★ ★

In 1823 James Mill moved to secure his son's future material well-being. Within four years of his appointment in 1819 as one of three Assistants in the Examiner's Office of the East India Company, James Mill had achieved a position in the Office second only to that of the Examiner himself. By April of 1823 his annual salary stood at a handsome £1200. His influence was such that in the following month he was able to obtain for his eldest son a junior clerkship in the Examiner's Office. The terms of this appointment stipulated that John Mill was to receive no salary during his first three years of service (the Company expressed its token appreciation in the form of an annual gratuity of £30). His initial appointment differed from that of other junior clerks in one important respect. He was to work under the supervision of his father and to be "employed from the beginning in preparing drafts of dispatches, and be thus trained as a successor to those who then filled the higher departments of the office."[78] James Mill was understandably pleased. The day after his son's appointment he wrote to his friend Dr. Thomas Thomson (author of *A System of Chemistry*): "The court of Directors have . . . appointed John . . . on a footing on which he will in all probability be in the receipt of a larger income at an early age than he would in any profession; and as he can still keep his hours as a student of law, his way to the legal profession is not barred, if he should afterwards prefer it."[79]

James Mill here seems ready to allow that the time might come when his son's preference, rather than his own, should determine John's career path. Yet this was more a concession in form than in substance. Employment at India House promised a degree of economic security that neither the law nor any other profession could assure. The fact that such employment did not rule out preparation for entry into the legal profession made it appear that the father had not irrevocably decided his son's career track. Suppose such employment had been thought incompatible with training for the bar—would James Mill then have kept his son away from India House? No. Would John Mill have refused to cooperate with his father had he concluded that in accepting a junior clerkship in the Examiner's Office he would thereby foreclose the prospect of a legal career? No. The East India Company had given James Mill the opportunity to gain a comfortable living while working to improve the quality of British governance in India. Placing a high value on his services, the Court of Directors had rewarded him accordingly. He had every expectation that his son would be treated with the same consideration. In theory John Mill's move into India House did not end the possibility of his qualifying for a career at the bar; in effect a decision had been made that a compliant John Mill would have to live with.

The somewhat perfunctory discussion in the *Autobiography* of his thirty-five year India House career indicates that it was not hard to live with; indeed, this discussion tends to confirm the wisdom of his father's judgment. Here J.S. Mill says that service in the East India Company "gained" him a "subsistence," which is to understate the case.[80] In 1825 he received a gratuity of £100, a figure that in the following year rose to £200; his appointment in 1828 to the position of Fourth Assistant to the Examiner gave him an annual salary of £310 (in 1828 John Mill was twenty-two years of age); his appointment as First Assistant in 1836 raised his remuneration to £1200 per annum.[81] He had a ready aptitude for the work, which generally stimulated his mind without taxing it. The job amply met his material wants without frustrating his wish "to devote a part of the twenty-four hours to private intellectual pursuits."[82] His duties, he remarks in a passage that his wife persuaded him to remove from the text, "occupied fewer hours of the day than almost any business or profession . . . had nothing in them to produce anxiety, or to keep the mind intent on them at any time but when directly engaged in them."[83] (Mill does not feel called upon to tell his readers that a large portion of both *A System of Logic* and *The Principles of Political Economy* was written at India House on India House stationery.) The experience gained from execution of his official responsibilities, he observes, also taught him much about the opportunities and constraints inherent in the "practical conduct of public affairs." What he learned at India House apropos of the conditions bearing upon the transaction of public business proved valuable to a "theoretical reformer of the opinions and institutions" of his age.[84]

Mill acknowledges that his position in the East India Company had "drawbacks, for every mode of life has its drawbacks." If it chimed well with "private intellectual pursuits," the same could not be said for its harmony with public political pursuits. Those who answered to the Court of Directors of the East India Company were not eligible for a seat in Parliament, an "exclusion," Mill says in the *Autobiography*, to which he "was not indifferent."[85] The restrictions on public forms of political participation did not end there. When Bentham acted in 1823 to fund the creation of a radical organ to rival the influence of the Whig *Edinburgh Review* and the Tory *Quarterly Review*, he apparently did not grasp that James Mill's employment at India House disqualified him from the editorship of the new periodical. According to John Mill, Bentham made the offer, which the elder Mill had to refuse.[86] The seventeen-year-old who became an apprentice to his father in the Examiner's Office in 1823 knew that certain ambitions of a political character could not be reconciled with service in the East India Company.

Whether he accepted, at this juncture, that a definitive choice had been made about his future employment is not knowable. Probably he did not seriously ponder the matter during the early years of his service. He no doubt took it for granted that the official work done by his father was work of some consequence. He understood that he was being trained to perform the tasks his father performed, and that he alone of the junior clerks appointed at this time was being so trained. The arrangement signified James Mill's confidence that his son could acquit himself well in the company of experienced men accustomed to discharging official functions. Awareness of this unusual status may well have been a source of satisfaction to John. His commitments at India House in the early and mid-1820s plainly did not interfere with his taking on myriad projects in which he was keenly interested and that had no connection with his place of employment. Who could tell what exploitable opportunities might yet present themselves to a young man of such outstanding ability and grand ambition? The challenge was to think in such terms while handling the continued dominion of his father, the father to whom he owed so much.

For John Mill the chief immediate "drawback" to his assuming a place at India House was that it expanded his father's direct jurisdiction over him. From this jurisdiction he had been temporarily liberated during his stay in France. Once back in England he had of course returned to his father's house. James Mill's India House responsibilities, however, now kept him away from home many hours each week, and his son, although expected to give instruction to his siblings, had a good deal more control over his time, and space, than had been the case before he left for France. With his appointment at India House in 1823, that control was in some considerable measure reduced.

The strain arising from this change was evidently not compounded by any expression of displeasure on James Mill's part regarding the quality of his son's work at India House. The younger Mill gave repeated demonstration of an intellectual maturity well beyond his years. More was not asked of him than he was able to deliver. That he met the expectations of the person to whom he was most accountable, and earned the respect of all those who reviewed his work, is implicit in what he says in the *Autobiography*. "My drafts of course required, for some time, much revision from immediate superiors, but I soon became well acquainted with the business, and by my father's instructions and the general growth of my own powers, I was in a few years qualified to be, and practically was, the chief conductor of the correspondence with India in one of the leading departments, that of the Native States."[87] James Mill's position within the East India Company helped make possible his son's rapid rise; John's performance made a reality of that possibility.

★ ★ ★

John Mill's opportunities to exercise his exceptional abilities at this time were not confined to the offices of the East India Company. The manifold activities he then engaged in—the discussion groups and debating societies, the contributions to radical newspapers and periodicals, the editing of Bentham's *Rationale of Judicial Evidence*—provided outlets for his emergent aspirations to participate vigorously in "the march of mind," a march both father and son were eager to press forward. The manner and substance of John Mill's early journalism and debating speeches abundantly display the attitudes, assumptions, and governing ideas of his mentors. Be the subject human nature, the proper mode of reasoning, political institutions and representation, the deformities of the legal system, freedom of the press, religious persecution, the pernicious influence of "sinister interests" (the aristocracy in particular), the deficiencies of the universities, or the tenets of political economy, the treatment given by the adolescent Mill was fully in accord with the doctrines propagated by his father and Jeremy Bentham.[88] In the opinions he expresses he appears to have complete confidence; in the tone, the example of Charles Austin is often discernible. "Judges, like other men, will always prefer themselves to their neighbours. Judges, like other men, will indulge their indolence and satiate their rapacity whenever they can do it without fear of detection."[89] What John Mill said and did in the five years after his return from France shows no sign of dissatisfaction with the intellectual edifice constructed by Bentham and his father.

The young man's search for companionship among his peers did, however, create tensions with his father. James Mill had striven to control his eldest son's social contacts. In the early 1820s this became increasingly difficult. As already noted, the decision to have John Mill pursue legal studies under John Austin had led to his acquaintance with Charles Austin and his Cambridge friends. These men were some years older than John Mill, and their circumstances were unlikely to bring them into close association with him (although he may have seen "much" of Charles Austin for a "year or two" during the early 1820s, nothing Mill says in the *Autobiography* implies that they became "friends" in any meaningful sense). The case of John Arthur Roebuck was altogether different. The death of Roebuck's father, an Indian civil servant, had been followed by the family's move to Canada, where Roebuck had been raised. Arriving in England in 1824 to study for the bar, the twenty-two-year-old Roebuck looked up Thomas Love Peacock, with whom the family had a connection. Peacock was an Assistant Examiner at India House, and he directly introduced Roebuck to the younger Mill. They struck up a fast friendship. Possessing a strong mind but little formal education, Roebuck was inducted into Philosophic Radicalism by John Mill.[90] Another young man who had fallen under John Mill's sway was George John Graham, a member of the Utilitarian Society formed in 1822–23. Graham too became close to Roebuck, and both would routinely accompany Mill on his morning walk to India House.[91] When John Mill brought Roebuck and Graham for a Sunday visit to Dorking, where James Mill rented a house for several months each year, his father took no trouble to conceal his dislike for his son's companions. He had words with Roebuck, a combative personality with a large ego and immense self-confidence. Roebuck and Graham returned to London on Monday morning, John briefly remaining behind to express his displeasure with his father's conduct. Upon "next seeing his friends, he told them what happened between him and his father; he had, he said, 'vindicated his position.'" According to Alexander Bain, the "scene left a great impression in the family."[92] John Mill kept up his friendship with Roebuck and Graham, but he did not again take them where they clearly were not welcome.[93]

The extraordinary nature of this incident naturally produced a lasting impression on those who experienced it, including all members of the Mill household. It marks the first reported instance of John Mill's openly challenging the judgment and behavior of his father. In defense of friendships he valued, and as a matter of personal honor, John Mill stood up to his father, an act requiring no small amount of courage. He did so because he had to show himself and his two friends that he was "a man among men," even if this entailed a quarrel with his father. The first reported instance of such frank opposition would also be the last.

★ ★ ★

We know of Mill's mental crisis from one source only, the *Autobiography*. No documents contemporary with the crisis attest to its existence. If friends or family noticed anything odd, they made no mention of it in any written form that has survived. Mill states that there was no one in whom he could confide, and implies that his demeanor would not have alerted those around him to his predicament. "During this time [the winter of 1826–27] I was not incapable of my usual occupations. I went on with them mechanically, by the mere force of habit."[94] Mill depicts the crisis in the pivotal chapter of the *Autobiography*, titled "A Crisis in My Mental History. One Stage Onward." In Mill's account the crisis ushered in a critical stage in his emotional and intellectual development. Although a number of elements define this new stage, its most salient feature is that John Mill, not James Mill, became the directing force.

The passage describing the onset of Mill's depression is among the best known in the literature of nineteenth-century autobiography.

> It was in the autumn of 1826. I was in a dull state of nerves, such as everybody is occasionally liable to; unsusceptible to enjoyment or pleasurable excitement; one of those moods when what is pleasure at other times, becomes insipid or indifferent; the state, I should think, in which converts to Methodism usually are, when smitten by their first "conviction of sin." In this frame of mind it occurred to me to put the question directly to myself, "Suppose that all your objects in life were realized; that all the changes in institutions and opinions which you are looking forward to, could be completely effected at this very instant: would this be a great joy and happiness to you?" And an irrepressible self-consciousness distinctly answered, "No!" At this my heart sank within me; the whole foundation on which my life was constructed fell down. All my happiness was to

have been found in the continual pursuit of this end. The end had ceased to charm, and how could there ever again be any interest in the means? I seemed to have nothing left to live for.[95]

Mill says there was nothing notable about the "dull state of nerves, such as everybody is occasionally liable to." What made a commonplace depression a crisis with profound repercussions was his posing of the question and the nature of the answer that followed from it. Both Alexander Bain and Leslie Stephen, among the first to comment on the mental crisis in the context of a biographical treatment of J.S. Mill, find the explanation for the depression in the effects of "overwork." Later commentators, whatever position they adopt on the core causes of the crisis, generally concede that such effects made Mill more susceptible to low spirits in the autumn of 1826 than he might otherwise have been. His activities in 1825–26 were enough to tire the most resourceful of young men. The editing of Bentham's *Rationale of Judicial Evidence*, from a disorderly series of manuscripts legible only to the most practiced eye, alone was a task of gigantic proportions.[96] The strain of such labors may have rendered Mill more vulnerable to a depression whose initial characteristics were perfectly ordinary; it cannot explain why a transfiguring crisis ensued.

The significance of Mill's strenuous efforts during the months preceding his mental turmoil had less to do with fatigue than with a rising awareness of his altered relation to his father. John Mill began to sense in the autumn of 1826 that the man from whom he had learned so much had little left to teach him. By this time the younger Mill had proven himself at India House. His appointment as a salaried clerk had been approved in May 1826. Even before this happened he had been moved to the newly established Correspondence Branch, and it had "become obvious to his clerical colleagues that the young man's promotion over their heads to the Examiner class in the Correspondence Branch was only a matter of time."[97] He had completed the editing of Bentham's *Rationale*, the five hefty volumes of which were making their way through the press. Especially noteworthy is what John Mill says in the *Autobiography* about the articles he wrote for the *Parliamentary History and Review* in 1825 and 1826. The topics he addressed—Catholic disabilities and the suppression of Daniel O'Connell's Catholic Association, the commercial crash of 1825–26, and the principle of reciprocity in international trade—derived from the issues of the day taken up by Parliament. They did not spring from the head of James Mill. His essays for the *Parliamentary History and Review*, John Mill observes, "were no longer mere reproductions and applications of the doctrines I had been taught; they were original thinking, as far as that name can be applied to old ideas in new forms and connexions: and I do not exceed the truth in saying that there was a maturity, and a well-digested character about them, which there had not been in any of my previous performances."[98] By the autumn of 1826 John Mill had learned how to function as a powerful and independent intellectual agent. On this score he had no further need of his father. Not only could James Mill not provide what his eldest son still did need, a centered and vibrant emotional life, but his treatment of his son had hampered the formation of such a life.

John Mill's mental crisis stands as a kind of ritual marking of his separation from his father and his embarking upon a new internal course of his own devising. Its emblematic status is implicit in the *Autobiography*. Explaining why he could not turn to his father for help, he says that his education had been "wholly" the work of James Mill, who had "conducted" it "without regard to the possibility of its ending in this result"; the predicament in which he now found himself was "beyond the power of *his* [James Mill's] remedies."[99] The training he had received from his father had given disproportionate attention to the expansion of his analytical powers while neglecting his affective endowments. Essential as analysis was to the discovery of truth, its habitual exercise, when coupled with the habitual slighting of other mental faculties, tends "to wear away the feelings."[100] His education had taught him to associate pleasure with the promoting of the general interest, and pain with the thwarting of that interest. It had not given him the means to build an emotional investment in the moral ends his reason had been schooled to fix upon. When his sagging spirits led him to ask what these ends meant to him, the answer made him desolate. With misery, however, also came some understanding of his plight. The antidote to despair would have to be found within himself.

Much has understandably been made of the event to which John Mill ascribed the beginning of his escape from this wretchedness. While reading, "accidentally" he says, Marmontel's *Memoirs*, he "came to the passage which relates his father's death, the distressed position of the family, and the sudden inspiration by which he, then a mere boy, felt and made them feel that he would be everything to them—would supply the place of all that they had lost." He wept. "From this moment my burthen grew lighter. The oppression of the thought that all feeling was dead within me, was gone. I was no longer hopeless: I was not a stock or a stone. I had still, it seemed, some of the material out of which all worth of character, and all capacity for happiness, are made."[101] Although John Mill held a kind of copyright on the presentation of his mental crisis, this aspect of his presentation unwittingly handed twentieth-century commentators with a psychoanalytical bent the key to deciphering it.[102] The nub of the crisis was the son's wish for the father's annihilation. Such an interpretation oversimplifies the complex cluster of feelings the younger Mill had for his father. Allowing that a sublimated desire for James Mill's final exit may well have been in play, it is nonetheless rash to give this desire precedence in an understanding of the mental crisis. The jumble of sentiments his father engendered in John Mill should caution against such a reductive explanation. If much is to be made of his account of the scene in Marmontel's *Memoirs* that inaugurated the recovery phase of the crisis, the profound sense of loss attendant upon the father's death should not be overlooked. A revival of feeling, in John Mill's case, was likely to be linked with the representation of strong personal feeling on the printed page. The "death-wish" strand should not be isolated from the tangled skein of emotions responsible for his sympathetic identification with the boy and his family.

In 1826 John Mill was no longer "a mere boy." He was an accomplished young man who had become acutely aware that his own life's plan had hitherto been devised by his father. His father's "plan"—one that neither John Mill nor any other child could possibly have formed for himself—had in most respects proved notably satisfactory. Its inadequacy did not lie in its observable attributes: the rigorous intellectual training, the apprenticeship in utilitarian radicalism, the position at India House. For all this, John Mill had cause to be grateful, and grateful he was. What the plan lacked was provision for an internal culture of the "feelings." The problem faced by the younger Mill in 1826 involved what to make of what he had been given, and how to give himself the emotional nourishment his father had failed to supply. Answering the latter challenge could go some way toward answering the former. His task was to complement and complete the superb, and indispensable, analytical education he had received. Between what he had been taught, and what he could teach himself by tilling an internal region to which James Mill had no access, he could create something valuable of his own. This undertaking would constitute an assertion of independence, even if his father could not be expected to grasp that such an assertion was being made. If John Mill was to go on with his life, this was the only helpful course open to him. James Mill retained a commanding presence in a large part of the physical space occupied by his son. Circumstances and temperament conspired to enforce upon John Mill a reticence that itself was symptomatic of his predicament. Without confronting his father directly, John would claim the measure of autonomy he needed by embarking on a quiet and intensive program of self-cultivation.

In the *Autobiography* Mill characterizes this program as one of the two far-reaching effects of the mental crisis and its resolution. He speaks of "the internal culture of the individual," of "the cultivation of the feelings," which he now saw as "one of the cardinal points in my ethical and philosophical creed."[103] The experience of reading Wordsworth's poems figures significantly in Mill's account of his recovery. For him they embodied "the very culture of the feelings, which I was in quest of. In them I seemed to draw from a source of inward joy, of sympathetic and imaginative pleasure, which could be shared in by all human beings."[104] To what extent had Mill's education impeded the emergence of an "internal culture" of the kind he extols here? He gives the impression that before his descent into depression he had assigned "almost exclusive importance to the ordering of outward circumstances, and the training of the human being for speculation and action."[105] His own training had indisputably placed a premium on inquiry, analysis, and action. This is not to say, however, that it failed to make room for anything else. Leaving aside the numerous histories and travel literature that had stimulated his youthful imagination, he had also read the following before his eighth birthday: *Robinson Crusoe*, Edward Forster's translation of the *Arabian Nights*, Jacques Cazotte's translation of *Arabian Tales*, Tobias Smollett's translation of *Don Quixote*, and Maria Edgeworth's *Popular Tales*. Some of these works James Mill had borrowed for his eldest son's entertainment. His "father's system," J.S. Mill notes in the *Autobiography*, did not "exclude books of amusement, though he allowed them very sparingly."[106] The mature John Mill had no reason to see his reading of such books as helping to prepare the way for his embrace of Wordsworth. Yet we should be wary of the idea that his training had stifled his imagination, especially in light of the vivacity of his response to the ancients, to historical literature, to France and the French Revolution, to music ("the only one of the imaginative arts in which I had from childhood taken great pleasure"), and to natural landscapes.[107] As for the last, he says explicitly that "the power of rural beauty over me" had created the groundwork for the pleasure he discovered in Wordsworth's poetry.[108] The sensibility that predisposed John Mill to find solace where he did, while not deliberately fostered by his father, had evolved contemporaneously with the education he had received from James Mill. He now considered a *policy* of internal culture essential to his well-being, and the crystallizing of this conviction no doubt did constitute an important change.

The other major change arising from this period of self-examination, we learn from the *Autobiography*, was John Mill's realization that the pursuit of personal happiness was incompatible with its attainment. Enjoyment could be derived from the passing pleasures life offers, but to make such enjoyment the predominant purpose of life was self-defeating. "Ask yourself whether you are happy, and you cease to be so. The only chance [for happiness] is to treat, not happiness, but some end external to it, as the purpose of life."[109] The external ends he mentions, by way of example, are "the happiness of others," "the improvement of mankind," and an "art or pursuit, followed not as a means, but as itself an ideal end."[110] At least the second of these examples would almost certainly have been on James Mill's list as well, and if John Mill ever believed that a quest for personal happiness could yield satisfactory results he could not have got this notion from a father who "had . . . scarcely any belief in pleasure."[111] The seasoned J.S. Mill says that he "never . . . wavered in his conviction that happiness is the test of all rules of conduct, and the end of life."[112] Of his father's moral standard, he had said much the same thing.[113] Although the "theory of life" John Mill now adopted avidly accommodated pursuits less austere than those valued by James Mill, the foundational elements of the theory were rooted in his education.

Unlike the above investigation of John Mill's mental crisis, his own commentary on the episode makes scant reference to his father. The education is there; the educator is largely absent. At its heart this experience was a self-enactment that disclosed John Mill's beliefs about himself. It could not be "shared" with James Mill, for it belonged entirely to the son, who now claimed sole responsibility for what he was to become.

CHAPTER TWO

Gathering Truths, 1826–30

As John Mill emerged from the worst of his mental crisis, he embarked on a search to discover truths that his education had failed to disclose. He sought to broaden the scope of his intellectual and emotional sympathies and to deepen their imprint upon his being. To this undertaking he brought several important advantages, not all of which he necessarily recognized. Endowed with a spacious, supple, and prodigiously powerful mind, John Mill had experienced an education that furnished him with a rich store of knowledge on an impressive range of subjects. To this knowledge was joined a command and appreciation of ordering principles that helped give shape and method to his quest. Moreover, his internal ordeal of 1826–27 forced him to confront his shortcomings and produced a vow to rectify them. And this was not all. John Mill had a secure position at India House, one that afforded him ample opportunity to pursue his program of internal culture without undue distraction. Whatever anxieties continued to beset him, they should not have arisen from concern for his present or future material well-being. As for the prospects of enriching the content of his emotional life, they were favorable. Cultivating the sympathies comes more easily when one is the object of others' affection and sympathy. John Mill found such affection and sympathy in a number of quarters. For the gathering of truths—about the world and about himself—he was favorably situated.

★ ★ ★

The physical realities of John Mill's existence in the late 1820s were, in the main, benign. By this time James Mill's family enjoyed a notable measure of material comfort. When appointed Assistant Examiner in

April 1823, the elder Mill's annual salary rose to £1200. When the younger Mill became Fourth Assistant to the Examiner, in February 1828, his starting salary, not including gratuities, was £310. A family income of this size placed the Mills firmly within the ranks of the upper middle classes. The wartime income tax introduced by William Pitt in 1799, from which the laboring poor were meant to be excluded, did not apply to incomes below £60; a revision of 1806 reduced the base limit to £50. Skilled London craftsmen, and constables in the Metropolitan Police Force created by Robert Peel in 1829, could expect to earn this much. Although a modicum of material well-being could be supported on £50 a year, comfort of a more substantial sort usually presupposed an income of £100 or higher. For the year 1815 some 160,000 persons were charged under Schedule D, which pertained to income derived from trade, industry, and professional activity. Nearly 60 percent of these reported incomes of less than £100. Only 12 percent reported incomes of £300 or more. Fewer than 5000 individuals reported incomes above £1000.[1] Before the end of 1830 James Mill's annual salary stood at £1900 (in this year he was promoted to Examiner). Upon this salary no direct tax was imposed, the income tax having been repealed by Parliament in 1816 (it would not be reintroduced until 1842, six years after James Mill's death). An income of £1900 exceeded the annual rentals collected by a significant portion of England's landed gentry.[2] Ultimately, James Mill had made very handsome provision for himself and his family.

The Mills retained their residence in Queen Square until the spring of 1831, when James Mill moved his family to Vicarage Place, Church Street, Kensington, to occupy a large detached villa. Three years before this move James Mill had acquired a summer residence at Mickleham in Surrey, a three-hour coach journey from London. Thomas Carlyle, who paid a visit to his friend John Mill when the latter was at Mickleham, liked the environs and the house. "It is a pretty country, a pretty village, of the English straggling wooded sort: the Mills have joined some 'old Carpenter's-shops' together, and made a pleasant summer mansion (connected by shed-roofed passages), the little drawing-room door of glass looking out into a rose-lawn, into green plains, and half-a-mile off to a most respectable wooded-and-open broad-shouldered Green Hill."[3]

Although he had an income sufficient for independence, John Mill continued to live at home with his parents and siblings, the education of the latter remaining his responsibility. His practical domestic needs were largely met by others—his mother, assisted by her older daughters, and the Mills' household servants. In a material sense, he lacked for nothing.

His place of occupation provided a generally congenial setting for labors not excessively taxing. The East India House, a splendid stately structure with grand Doric columns, had been rebuilt by the proprietors of the Company during the late 1790s. It stood majestically in Leadenhall Street in the City. True, James Mill had a magisterial presence at India House, not an unmixed blessing to the younger Mill. In a practical sense, however, the father's expertise helped the son acquire, in fairly short order, a self-directed proficiency over the business that came his way. John Mill's demonstrable ability won him promotion to the rank of Assistant to the Examiner in 1828.

The income and way of life of James and John Stuart Mill indicate they belonged to the upper portion of the "middle rank" of English society. Yet their incomes had no direct connection with either the established professions of early-nineteenth-century England—the church, the officer class of the army and navy, the law, and medicine—or with the worlds of commerce, investment, and industry (the empire of the East India Company had as much to do with conquest and control of land revenue as with anything resembling conventional commercial activity). Their positions at India House, practically unaffected by market forces, insulated the Mills from the acutely competitive and precarious universe in which most members of the English middle classes dwelt. W.J. Fox, a prominent Unitarian minister, radical journalist, and editor, stated in 1835: "in the middle classes we note an almost universal unfixedness of position. Every man is rising or falling or hoping that he shall rise, or fearing that he shall sink."[4] Five years later John Mill himself noted the "entire unfixedness in the social position of individuals" within England's middle classes, citing "that treading upon the heels of one another—that habitual dissatisfaction of each with the position he occupies, and eager desire to push himself into the next above it."[5] At one time "unfixedness" had been the lot of James Mill, but not after 1819 and his appointment at India House. John Mill was spared the struggle his father had endured. The status and security he enjoyed by virtue of the service he gave the East India Company did not prevent John Mill from striving to do great things. They allowed him to focus this striving outside the economic arena.

John Mill had no more reason to feel insecure about his personal attributes than he did about his professional circumstances. His bodily constitution, if not especially robust, had the soundness necessary to sustain good health during the late 1820s and early 1830s. Nothing suggests he was at all vain about his person, yet he had no need to worry that others would be put off by his presence. Above average in height, slim, fair-haired, ruddy-complexioned, and possessing an attractive face, Mill's form and features made a vivid and pleasing impression. Thomas Carlyle first met Mill in early September 1831, and soon after told his wife Jane of this "slender rather tall and elegant youth."[6] Caroline Fox, member of a notable Quaker family, first saw John Mill in March 1840, by which time he had aged noticeably; she found "his exquisitely chiseled countenance . . . beautiful and refined."[7] Henry Taylor, a contemporary of the younger Mill and a fellow-member of the London Debating Society, described his manners as "plain, neither graceful nor awkward; his features refined and regular; the eyes small relatively to the scale of the face, the jaw large, the nose straight and finely shaped, the lips thin and compressed, the forehead and head capacious; and both face and body seemed to represent outwardly the inflexibility of the inner man."[8] Good as it was for the purposes required of it, Mill's body did not perform most physical tasks with ease. He himself noted that he never got over being "inexpert in anything requiring manual dexterity."[9] This clumsiness did not, however, interfere with his passion for walking, and his way of life was such that it did not prove a serious liability.

This attractive young man had companions with whom to walk and talk. People liked John Mill, and he showed a sociable disposition in these years. His father did not see much of this geniality, but other family members evidently did. In 1830 James Mill's second son, James Bentham Mill (born 1814) entered University College, London. A classmate, Henry Solly, spent a week with the family at their Surrey cottage. After John Mill's death, Solly wrote up his impressions of this visit in two separate accounts. In one of these he remarked on "the affectionate playfulness" of John Mill's "character as a brother in the company of his sisters, and of the numerous younger branches of the family."[10] In the other he noted: "John Mill always seemed to me a great favourite with his family. He was evidently very fond of his mother and sisters, and they of him; and he frequently manifested a sunny brightness and gaiety of heart and behaviour which were singularly fascinating."[11]

John Mill's sociability extended well beyond the domestic sphere. He frequently joined other guests at weekly breakfasts hosted by Henry Taylor (habitual attendees included Charles Austin, John Romilly, Edward Strutt, and Charles Villiers).[12] In his *Autobiography* Taylor wistfully recalled that these gatherings, which began at ten o'clock, sometimes did not end before three.[13] On 15 November 1830 Charles Greville, clerk to the privy council—and the most penetrating political diarist of his age—met John Mill, among others, at one of Taylor's breakfasts. "Young Mill," he recorded in his diary, "is Son of Mill who wrote the 'History of British India,' and said to be cleverer than his father. He has written many excellent articles in reviews, pamphlets, etc., but though powerful with a pen in his hand, in conversation he has not the art of managing his ideas, and is consequently hesitating and slow, and has the appearance of being always working in his mind a proposition or syllogism."[14] The Mill Alexander Bain came to know in the early 1840s had a conversational efficiency unnoticed by Greville. Bain affirmed that Mill "always aimed at saying the right thing clearly and shortly. He was perfectly fluent, but yet would pause for an instant to get the best word, or the neatest collocation: and he always liked to finish with an epigrammatic turn."[15] Carlyle met Mill less than a year after the repast at Taylor's to which Greville referred. In a letter to his wife, Carlyle spoke of Mill's remarkable gifts and "precision of utterance."[16] Henry Taylor himself implied that Mill held his own during the breakfast attended by Greville, which also included the Tory poet Robert Southey. Taylor alluded to the "odd assemblage" in a letter written the following day: "There were Southey and Mill, as far as the poles asunder in politics, but somewhat akin in morals and habits of literary industry." All members of the party, he observed, "talked copiously and well."[17] John Mill's pen being more forceful than his tongue did not keep Taylor from seeing the power in both.[18]

Another "Henry" saw a lot of Mill in the late 1820s. Henry Cole, Mill's junior by two years, lived with his father in a house that belonged to Thomas Love Peacock, James Mill's colleague (and a comic novelist of note). On 7 November 1826 Peacock brought Cole and Mill together. The entry in Cole's diary for this date tersely states: "Introduced by Mr. Peacock to Mr. John Mill."[19] Thereafter Cole subscribed to the London Debating Society and sometimes joined Mill, Roebuck, and Graham in the small discussion group that met twice a week at the home of George and Harriet Grote in Threadneedle Street. Cole often stopped in at India House to visit Peacock, Horace Grant, and John Mill before moving on to the Tower of London, where he carried out duties connected with his post on the Records Commission.[20] A number of autumn evenings in 1828 found Cole taking tea with John Mill in Queen Square. On 4 September of this year Mill showed Cole his collection of "Botanical Specimens," a number of which he presented to his friend as a gift.[21] Cole graduated from plants to people on 19 November, when he was introduced to Mill's "Mother & Sisters— without prejudice & agreably [sic] well educated personages."[22] Agreeable as John Mill might have found the company of his mother and sisters during these years, they could not compete with a flower he unexpectedly encountered in June 1830. Cole's diary recorded his friend's rapture: "Called on John Mill who was exulting in his discovery of the Martigon Lily at Dorking."[23]

The friendship of Cole and Mill encompassed conversation, tea, flora, walks, and music. Mill mentioned, in an early version of the "Early Draft," that in late winter and early spring he would, weather permitting, "make a walking excursion with some of the young men who were my companions; generally walking out ten or twelve miles to breakfast, and making a circuit of fourteen or fifteen more before getting back to town."[24] Cole was one of these "young men," his diary for 1829 offering some particulars regarding these Sunday outings. On 19 April, for example, Cole "Walked with John Mill and Grant [Horace Grant, Mill's colleague and friend at India House] to Locks Bottom to Breakfast thro' Sydenham and Bromley, and retd [returned] by Orpington, Cheselhurst, & Sydenham."[25] The more Cole saw of Mill, the more he valued his friendship. In early February 1830, he noted in his diary: "Mill drank tea and passed the Evg with me. The admiration & esteem for his talents encreases [sic] each time that I have the pleasure of seeing him."[26] In the early 1830s Cole had a connection with two Millian walking tours—in Yorkshire and the Lake District for a month in the summer of 1831 (during which Mill visited Wordsworth), and in Hampshire, West Sussex, and the Isle of Wight for more than a fortnight in the summer following.[27] A shared enthusiasm for music also brought the two young men together. John Mill, a skillful amateur pianist who enjoyed playing the compositions of others and his own, found a musical companion in Cole. On Tuesday, 14 June 1831, "early in the morning," Mill called on Cole "to converse about Harmony." On Wednesday evening, 30 November 1831, Mill drank tea with Cole and "played a vast quantity of music from memory." Mill and Cole also attended the opera together.

Mill continued to be on cordial terms with many whose acquaintance he had made years earlier. Of those older than himself, the Grotes and the Austins retained a conspicuous place. In a personal sense, John Mill certainly felt closer to the Austins (their daughter Lucy included), whose company he would experience in a more intimate setting than that supplied by the Grotes. If he nonetheless saw the Grotes more often than he saw the Austins, this was simply because George Grote (a prosperous banker) had the resources, and Harriet Grote the inclination, to make their home a political and intellectual center for young radicals of a philosophic bent. The Austins manifestly lacked the means to act in this capacity during the late 1820s, years when they had to practice "the most rigid economy."[28] (Years later Sarah Austin became a notable literary hostess; her reclusive husband never took to the part of host.) During the second half of the 1820s the Grotes' residence served as the meeting place for the discussion group led by John Mill. Its members included Roebuck and G.J. Graham. Although the closeness he had felt with Roebuck in the mid-1820s had lessened by the end of the decade, John Mill certainly still considered him a "friend." Mill rated Graham's intellectual force above that of Roebuck, later describing Graham as "a thinker of originality and power on almost all abstract subjects."[29] Graham sometimes accompanied Mill on his Sunday walks.

Another contemporary Mill had known for some years, William Eyton Tooke, retained a place in his life in the late 1820s. Mill's *Autobiography* mentions Tooke several times: as "a young man of singular worth both moral and intellectual"; as an early contributor to the *Westminster Review*; as an agent in the spread of James Mill's influence among the generation of Cambridge undergraduates that followed Charles Austin's; and as the person through whom John Mill became acquainted with F.D. Maurice.[30] Mill's affectionate solicitude for Eyton Tooke comes through in a letter he wrote to the Grotes on September 1, 1824, a few days after he had been a dinner guest at the Wimbledon home of Thomas Tooke, Eyton's father. John Mill reported that Mr. Tooke had taken strong exception to some of the views expressed by his son during a lengthy discussion of economic questions. "Two or three times he put him down with considerable harshness," Mill observed. These "reproofs," Mill continued, the younger Tooke accepted "with the most infidel charity & resignation. I contrived, towards the close of the evening to take a turn with Eyton in the garden, and we had some very profitable conversation: he is eager to do whatever good he can, & to qualify himself for doing more."[31] Mill kept up his strong friendship with Tooke through the second half of the 1820s.

The social ties Mill maintained with such contemporaries as Tooke, Graham, Roebuck, Grant, Cole, and Taylor refreshed his spirits and exercised his sympathies. They gave to his routines a salutary balance and made him know that others valued his companionship. Missing, however, was a conviction that he could learn much from these young men. He liked and respected them (of Tooke he was especially fond), but did not find in any of them a mental power comparable to his own. They could not aid his on-going reappraisal of his intellectual heritage. In the mid-1820s he met two men who did make a significant contribution to this reassessment, Frederick Denison Maurice and John Sterling.

Mill met Maurice through Eyton Tooke, who had been at Cambridge during Maurice's years there. Maurice and Sterling drew close while at university, where both men had made names for themselves at the Cambridge Union and become early members of the "Apostles," a small coterie of intellectually gifted and morally earnest undergraduates. Like Tooke, Maurice and Sterling were almost exact contemporaries of John Mill. From 1827 to 1829, they brought to the London Debating Society a stimulating new dimension. As Mill says in the *Autobiography*, the Philosophic Radicals in the society now found themselves facing adversaries "of far greater intrinsic worth" than the Tories "with whom we had hitherto been combating."[32] Maurice and Sterling, John Mill observed, "were of considerable use to my development."[33] They "made their appearance in the [London Debating] Society as a second liberal and even Radical party, on totally different grounds from Benthamism and vehemently opposed to it; bringing into these discussions the general doctrines and modes of thought of the European reaction against the philosophy of the eighteenth century."[34] Mill saw Samuel Taylor Coleridge as the chief representative of this "reaction in England"; Coleridge's thought had profoundly influenced Maurice and Sterling.

They were probably introduced to Coleridge's work by Julius Charles Hare, then a young Cambridge don and Sterling's tutor at Trinity College. In 1818 Coleridge brought out a three-volume revised edition of *The Friend*. This collection of writings, originally composed for a periodical of the same title produced in 1809–10, included a piece of considerable length entitled "Essay on the Principles of Method." The central question explored by Coleridge in this largely new essay was what it meant to be an educated man. A passage toward the close of the essay resonated powerfully with a number of young men who later became part of early- and mid-Victorian England's "learned" classes. Coleridge referred to the "many examples . . . of young men the most anxiously and expensively be-schoolmastered, be-tutored, be-lectured, any thing but *educated*; who have received arms and ammunition, instead of skill, strength, and courage; varnished rather than polished; perilously over-civilized, and most pitiably uncultivated!" This condition stemmed "from inattention to the method dictated by nature herself, to the simple truth, that as the forms in all organized existence, so must all true and living knowledge proceed from within; that it may be trained, supported, fed, excited, but can never be infused or impressed."[35] This message imbued Hare's guidance of the Cambridge undergraduates he took under his wing. Hare greatly admired Wordsworth and Coleridge, both of whom he counted among his friends; he brought Maurice and Sterling into Coleridge's company.[36] Hare, according to his modern biographer, "reserved his worst venom for Jeremy Bentham and his followers."[37] Maurice and Sterling vigorously and ably fought the Benthamites in battles sponsored by the London Debating Society.

Mill was especially drawn to Sterling. The two joined in a defense of Wordsworth in a debate of January 1829. The proposition contested in the debate, "That Wordsworth was a greater poet than Byron," divided Mill from Roebuck—an enthusiast for Byron's poetry—and made allies of Mill and Sterling. Sterling opened the debate, with what Henry Cole described in his diary as "a long and rambling speech." He was answered by Roebuck, who, in "a most excellent speech," put forward "a good case for Byron." Mill proved Roebuck's most formidable foe in a speech that lasted two hours. Cole considered it "a most excellent essay"; Sterling thought it "admirable," and "infinitely better" than his own effort. Only in conversation with Coleridge, Wordsworth, and Maurice had Sterling ever "seen or heard anything like the same quantity of acute & profound poetical criticism."[38] Sterling also could not have missed the personal turn Mill gave to his appreciation of Wordsworth's merits.

> I have learned from Wordsworth that it is possible by dwelling on certain ideas to keep up a constant freshness in the emotions which objects excite and which else they would cease to excite as we grew older—to connect cheerful and joyous states of mind with almost every object, to make every thing speak to us of our own enjoyments or those of other sentient beings, and to multiply ourselves as it were in the enjoyments of other creatures: to make the good parts of human nature afford us more pleasure than the bad parts afford us pain—and to rid ourselves entirely of all feelings of hatred or scorn for our fellow creatures.[39]

Decades later Mill remarked that in the wake of this debate Sterling confided that he had mistakenly regarded Mill as a "manufactured man." His discovery "that Wordsworth, and all which this name implies, 'belonged' to me as much as to him and his friends" had transformed his estimate of Mill.[40]

Mill would come to feel a keen affection for Sterling. A man of immense natural charm, endearing amiability, great sincerity, and high moral rectitude, Sterling evoked powerful feelings of attachment from those privileged to be his friends. Mill cherished his memory long after Sterling's death from consumption in 1844. Hare would compose a memoir, and Carlyle, who developed a robust fondness for Sterling in the years after Mill brought the two together, wrote a warm and admiring biography. Carlyle characterized this work as "a light portrait, the truest I could easily sketch, of an unimportant but very beautiful, pathetic and rather significant, human life in our century."[41] A passage from Mill's *Autobiography* recalls Sterling's genius for friendship.

> He was indeed one of the most loveable of men. His frank, cordial, affectionate and expansive character; a love of truth alike conspicuous in the highest things and the humblest; a generous and ardent nature which threw itself with impetuosity into the opinions it adopted, but was as eager to do justice to the doctrines and the men it was opposed to, as to make war on what it thought their errors; and an equal devotion to the two cardinal points of Liberty and Duty, formed a combination of qualities as attractive to me, as to all others who knew him as well as I did.[42]

Sterling's fine nature fastened upon Mill's sensibilities at a time when the latter was particularly receptive to new influences.[43]

★ ★ ★

In the second half of the 1820s John Mill showed a readiness to investigate the currents of European thought that he identified with the "reaction against the philosophy of the eighteenth century." The chief defect of Enlightenment thought, Mill reckoned, was its negativity. Although they possessed the invaluable merit of forcefully exposing the intolerance, superstition, prejudice, and falsehood that riddled the institutions and belief systems of the old order, the leading thinkers of the eighteenth century failed to understand the conditions shaping the rise of those institutions and systems of belief, or the essential role these had played in civilizing Europe. Nor had the best minds of the eighteenth century offered anything positive in place of what they aimed to destroy. They did not grasp "that when all the noxious weeds were once rooted out, the soil would stand in any need of tillage."[44] In contrast, "the Germano-Coleridgian school," which Mill saw as spearheading the reaction against the thought of the eighteenth century, "were the first who inquired systematically into the inductive laws of the existence and growth of human society."[45] His encounter with this school helped convince Mill that "the great danger to mankind is not from seeing what is not, but from overlooking what is."[46]

The impact of Maurice and Sterling presumably had something to do with Mill's "deriving much from Coleridge, and from the writings of

Goethe and other German authors" in the latter half of the 1820s.[47] Steeped in German literature and philosophy, Coleridge, in the words of his modern biographer, "championed the new German criticism and idealist philosophy, adapted it and developed it in an English context, and successfully made it part of the Romantic movement."[48] Unlike Coleridge, Mill never immersed himself in German thought and certainly never espoused the cause of German idealism. Learn German he did, in the mid-1820s, when he and a few of his friends "formed a class" to accomplish this end.[49] This pursuit, however, did not presage a systematic and sustained engagement with major German thinkers. Mill had but a superficial acquaintance with the thought of Immanuel Kant, the greatest German philosopher of the late eighteenth century, and a man whose work stood in an ambiguous relation to the main currents of the Enlightenment.[50] The same can be said of his familiarity with such German critics of the Enlightenment and Kantian rationalism as Johann Gottfried Herder, Friedrich Heinrich Jabobi, and Johann Georg Hamann. And what little Mill knew of Germany's early-nineteenth-century "philosophers of feeling," such as Franz von Baader, Friedrich Schleiermacher, and Jean Paul Richter, he mainly got from the writings of Coleridge and from a couple of early essays by Carlyle.[51] At no point did the specifically Germanic content of Coleridge's thought prove especially palatable to Mill.

Lack of evidence precludes a precise rendering of the nature and extent of Coleridge's influence on Mill before the end of the 1820s. Mill's publishing output in the late 1820s was meager, and the little he produced ignored Coleridge. The *Autobiography* offers scant help. In a section that sums up his state of mind respecting political institutions and principles circa 1830 he says: "The influences of European, that is to say, Continental, thought, and especially those of the reaction of the nineteenth century against the eighteenth, were now streaming in upon me. They came from various quarters: from the writings of Coleridge, which I had begun to read with interest even before the change in my opinions . . ."[52] The first reference to Coleridge in Mill's extant correspondence, in a letter to Sterling, does not appear until October 1831.[53] We do know from a speech he delivered in the London Debating Society in spring 1828 that Mill then already regarded Coleridge as one "of the wisest men of all political and religious opinions."[54] Mill's *Collected Works* include many references to Coleridge's poetry. Although these do not predate 1830, he surely must have been reading Coleridge as well as Wordsworth in the years immediately following the onset of his mental crisis. For the lifting of spirits, Wordsworth's poetry no doubt suited better than Coleridge's; as a poet whose verse fused imaginative power and intense feeling, Coleridge the artist nonetheless stood high in Mill's estimation, and played some part in the latter's affective education.

It was as a political philosopher rather than as a poet that Coleridge influenced Mill's thinking. Conceptual changes in Mill's view of political institutions and precepts make up the context for his point about having read "the writings of Coleridge . . . with interest." Of Coleridge's prose works, the two cited most often by Mill are the *Second Lay Sermon* and *On the Constitution of the Church and State*, the texts encompassing the central ideas of Coleridge's mature thought on politics and society. Two editions of the latter work carry a publication date of 1830, although the first appeared in December 1829.[55] When Mill invoked Coleridge as an example of political wisdom in his debating speech of 1828, he most probably had in mind the *Second Lay Sermon*, first published in 1817.

The *Second Lay Sermon* embodied Coleridge's response to the severe hardships afflicting the English masses in 1817. Dramatic cuts in government expenditure, the demobilization of the armed forces, high bread prices arising from a series of poor harvests, wage reductions, rising levels of unemployment and underemployment, and extremely rapid population growth combined to produce widespread distress. The discontent of the masses was expressed in the form of food riots, rick burning, and protest meetings. In the *Second Lay Sermon* Coleridge argued that the principal cause of the disordered state of English society was an excess of the commercial spirit. While acknowledging that commerce was an indispensable progressive force, Coleridge maintained that its spirit could not be allowed to dominate the social order. Selfish ends necessarily informed the pursuit of wealth, and these ends had to be balanced by forces identified with a sense of responsibility to the community. When functioning properly, the State itself, together with the agricultural order—whose leading members held most of the fixed personal property of the nation—incarnated values that promoted the durable interests of the wider society. Coleridge specified the ends they were to serve. "1. To make the means of subsistence more easy to each individual. 2. To secure to each of its members THE HOPE of bettering his own condition or that of his children. 3. The development of those faculties which are essential to his Humanity, i.e. to his rational and moral Being."[56] In Coleridge's judgment, these ends were not being met. Dismayed by the conduct of the government and the policies it had put forward, and disturbed by the unfeeling treatment shown tenants and agricultural laborers by the country's landlords, he attributed these regrettable developments to the commercial spirit's overspilling its legitimate bounds.[57]

By the time he came to write *On the Constitution of the Church and State*, Coleridge had decided that neither the government nor the leaders of landed society had the ability to fulfill the purposes set forth in his *Second Lay Sermon*. The "clerisy," as conceived by Coleridge, should have that ability. Coleridge regarded the Church of England as a National Church whose property was set apart for the use of national purposes. Its duty was to apply the resources it possessed to the education of the nation. This endeavor must go beyond the provision of rudimentary instruction. The education the clerisy imparted to the members of the community must be "grounded in *cultivation*, in the harmonious developement of those qualities and faculties that characterise our *humanity*. We must be men in order to be citizens."[58] Coleridge presented his *idea* of the National Church, and admitted that the condition of the actual Church of England in 1830 fell well short of realizing this idea. John Mill found much to admire in Coleridge's conception of what the National Church and the clerisy ought to be.

Mill read *On the Constitution of the Church and State* sometime before mid-autumn 1831. In a letter to Sterling dated 20 October 1831, he observed: "I certainly think it desirable . . . that there should be a national clergy or clerisy, like that of which Coleridge traces the outline, in his work on Church & State." Most of those who made up the clergy of the Church of England, he added, were unfit to discharge the functions of Coleridge's clerisy. Mill told Sterling that he believed Coleridge would agree with him (Mill) "in thinking that a national clergy ought to be so constituted as to include all who are capable of producing a beneficial effect on their age & country as teachers of the knowledge which fits people to perform their duties & exercise their rights, and as exhorters to the right performance & exercise of them."[59] Many of these capable individuals would not be members of the Church of England; some would not even be Christians. No manner of Christian himself, Mill certainly aspired to produce "a beneficial effect" on his "age & country." Coleridge's clerisy would exercise "the authority of the instructed," an authority that Mill was given to thinking a lot about during these years.

Another man giving it a lot of thought at this time was John Austin. In 1826 Austin was appointed Professor of Jurisprudence and the Law of Nations at the University of London, a new and undenominational institution of higher learning that James Mill had helped found.[60] Students were not expected to enroll in courses before 1828, and in 1827 the Austins left for Bonn, which offered an attractive setting, a low cost of living, and a university boasting five professors of jurisprudence. Here John Austin would study and prepare his lectures.[61] John Mill noted that this sojourn in Bonn "made a very perceptible change" in Austin's "views of life." He responded sympathetically to "the influences of German literature and of the German character and state of society." These observations appear in a section of the *Autobiography* that treats the late 1820s and early 1830s. Mill returns to Austin in this section because he, of "the persons of intellect whom I had known of old," was "the one with whom I had now many points of agreement." Like Mill himself, Austin's "tastes," colored by his German experience, had turned "towards the poetic and contemplative"; he now "attached much less importance than formerly to outward changes, unless accompanied by a better cultivation of the inward nature."[62] All this meshed well with Mill's unfolding concern for internal culture, with his rejection in principle of sectarianism and dogmatism, with his growing affinity for Goethe's ideal of "many-sidedness." Moreover, Mill plainly shared Austin's disdain for the selfishness, "meanness," and small-mindedness so prevalent in English society.[63] Although careful to distance himself from Austin's "indifference" to the advance of democratic institutions, Mill joined Austin in believing "that the real security for good government" rested in an instructed people.[64]

Austin delivered his first set of lectures in the autumn of 1829 (ill-health prevented his performing this duty in 1828).[65] John Mill attended these evening lectures (two each week), and compiled notes of what he heard. Some of the ideas Austin put forward would have an important bearing on Mill's subsequent exploration of the problem of authority in an age of transition.[66] In his third lecture, Austin urged that the systematic investigation of ethics subsumed "so spacious a field that none but the comparatively few, who study the science assiduously, can apply the principle [of utility] extensively to received or positive rules, and determine how far they accord with its genuine suggestions or dictates."[67] Austin and Coleridge differed on where to look for the ultimate source of moral truths; along with Mill himself, they were as one in seeing the need for "teachers of the knowledge which fits people to perform their duties & exercise their rights." While Austin conceded that a perfect theory and practice in the province of ethics could not be had, he nonetheless held that major advances were attainable. A commitment to impartial and patient inquiry, coupled with the application of systematic investigative procedures, "would thoroughly dispel the obscurity by which the science is clouded, and would clear it from most of its uncertainties."[68] The diffusion of greater knowledge and experience among the masses, Austin observed, would render them evermore inclined to accept the tuition of those highly instructed persons responsible for inculcating a compact and consistent body of doctrine; the deference of the multitude would be rooted in their reasoned respect for "the comparatively few" who possessed a deep and comprehensive understanding of moral science.

The state of things here envisaged by Austin appealed powerfully to Mill. The problem was finding the persons qualified to guide the masses as they should be guided. At the same time as he listened attentively to Austin, Mill came into contact with the Saint-Simonians, a school of thinkers who laid claim to the kind of authority Austin spoke of.

Of noble birth, Claude-Henri Saint-Simon (1760–1825) had renounced his title at the beginning of the French Revolution, for the first phases of which he showed some enthusiasm. Having survived the Terror, he served under the Directory. After giving up an active political life in the early years of the nineteenth century, he took up his pen in earnest and developed a body of social and political doctrine that would ultimately influence a number of important nineteenth-century thinkers. Especially noteworthy, in John Mill's terms, was Saint-Simon's notion of historical change. Advocates of social and political reform, Saint-Simon maintained, needed to understand the conditions governing such change. The destruction of obsolete institutions could be accomplished speedily enough, as the French Revolution had graphically demonstrated. The creation of durable and progressive institutions in their stead, on the other hand, required an astute analysis of the forces shaping historical development. Saint-Simon saw the collapse of the feudal order in Europe, and the transformations that followed, as bound up with conflicts between classes whose interests, attitudes, and aspirations were spawned by social, economic, and technological forces. The institutional order and belief system of medieval Europe had effectively expressed the organic stage of historical development that society had then reached. As shifts in the organization of economic production occurred in response to social, scientific, and technological change, medieval institutions and beliefs lost their social utility, and a critical stage of historical development began. A persistent erosion of the legitimacy of the old order eventually culminated in the era of the French Revolution. The dissolution of that order, however, had not yielded a new unity of purpose or the institutional framework essential to the achievement of a new organic state of society. Saint-Simon sought to furnish such a framework, the specifics of which Mill would find much less compelling than the conception of historical change in which Saint-Simon's practical recommendations were embedded.

By the time of Saint-Simon's death in 1825, his writings as a social philosopher had secured him a body of ardent disciples. One of these was Gustave d'Eichtal. The son of a wealthy banker, d'Eichtal was in his mid-twenties when he first met John Mill. In London to proselytize on behalf of the Saint-Simonian cause, he attended a gathering of the London Debating Society at the end of May 1828. Mill participated in the debate, and d'Eichtal, riveted by the intellectual force on display, resolved to win him over to Saint-Simonism.[69] He began a vigorous correspondence with Mill, and supplied him with tracts that set forth Saint-Simonian doctrine. Among the latter was Auguste Comte's *Systéme de politique positive*, published in 1824. Comte had become Saint-Simon's secretary in 1817, only to have an acrimonious falling out with the master some seven years later. Written before this breach, and avowedly Saint-Simonian in character, the work Comte brought out in 1824 came to Mill with d'Eichtal's hearty endorsement. Mill's response to the book, conveyed in a letter of October 1829, registers some of the effects of the changes set in motion by the crisis of the mid-1820s.[70]

At this time, Mill's admiration of Comte's capacity for expounding a system of thought did not extend to the method by which he arrived at his conclusions or to the content of the conclusions themselves. "M. Comte is an exceedingly clear and methodical writer, most agreeable in stile, and concatenates so well, that one is apt to mistake the perfect coherence and logical consistency of his system, for truth." Comte did not seem to grasp that the truths of politics and social science, unlike those of mathematics, could not be properly deduced "from a set of axioms & definitions." The method embraced by Comte necessarily produced an emphatically one-sided and incomplete account of what government and society were about. The "fundamental principle" of Comte's "whole system," Mill stated, is "that government and the social union exist for the purpose of concentrating and directing all the forces of society to some one end." Mill took strong exception to this notion, proclaiming that "Government exists for all purposes whatever that are for man's good: and the highest & most important of these purposes is the improvement of man himself as a moral and intelligent being, which is an end not included in M. Comte's category at all." Contending that it was neither possible nor desirable that one end alone should be pursued by the "united forces of society," Mill insisted that men "do not come into the world to fulfil one single end, and there is no single end which if fulfilled even in the most complete manner would make them happy."

When Mill turned to consider Comte's candidates for the "one single end," he found them unworthy of the precedence Comte assigned them. For Comte, the choice was between "the dominion of man over man, which is conquest, or the dominion of man over nature, which is production." Mill noted that Comte attributed the former end to the ancient world (Mill rejected this attribution) and the latter to the modern age. Had Comte witnessed at first-hand the consequences for English society of its attaching excessive importance to "production"—a preference that "lies at the root of all our worst vices, corrupts the measures of statesmen, the doctrines of philosophers & hardens the minds of our people so as to make it almost hopeless to inspire them with any elevation either of intellect or soul"—he might have drawn back from such a hazardous proposition. Mill also demurred from Comte's assertion that one law of development held sway over the course of civilization. Invoking England and France as examples of advanced civilizations, Mill observed that these two societies had followed different paths, and suggested that "neither of them has, nor probably ever will, pass through the state which the other is in." The faculties of man, he urged, were not subject to a uniform order of development, their formation displaying a variety as great "as the situations in which he is placed." Mill conceded that Comte's book contained "many excellent & new remarks"; when properly modified and corrected, these could prove "very valuable." It also contained a lot that was unsatisfactory. The followers of Saint-Simon should be alert to the work's inadequacies, and not treat its doctrines as articles of faith. If they acted as a dogmatic sect, he cautioned d'Eichtal, "they will not only do no good but I fear immense mischief." They should instead heed a dictum that Mill had recently adopted: "Substituting one fragment of the truth for another is not what is wanted, but combining them together so as to obtain as large a portion as possible of the whole."[71]

In a letter written a month later, Mill sought to clarify his assessment of "the Saint Simon school" by dilating upon what he approved and admired in its project. During the interval Mill had received several letters from d'Eichtal, for whom he had come to feel a genuine fondness. He did not want d'Eichtal to think that his efforts to gain Mill's sympathy and support had been fruitless, and worried that his previous "letter may have left an impression on your mind, that I do not think so highly of them [the Saint-Simonians] as I actually do." Mill proceeded to identify two distinguishing marks of Saint-Simonian thought that he considered of outstanding importance, the first being a fundamental belief in "the necessity of a *Pouvoir Spirituel*."[72] Comte, in articles printed in the Saint-Simonian periodical *Le Producteur* in 1825–26, had stipulated the need for an organized hierarchy of intellectual leaders, supported by the state, who would act to disseminate knowledge and regulate opinion within the wider society. Such a coterie would be indispensable for sustaining a cohesive new social order in the wake of the dissolution of the old.[73] The Saint-Simonians posited, in Mill's words, the achievement of "a state in which the body of the people, i.e. the uninstructed, shall entertain the same feelings of deference & submission to the authority of the instructed, in morals and politics, as they at present do in the physical sciences." In principle, Mill did not doubt that such was "the only wholesome state of the human mind." The rub lay in the absence of a qualified group of instructors, and in the presence of powerful sinister interests determined to prevent the formation of such an intellectual influence. These formidable hurdles aside, Mill had grave reservations about the means advocated by the Saint-Simonians "for organizing the *pouvoir spirituel*." Indeed, he submitted that "you cannot organize it at all. What is the *pouvoir spirituel* but the insensible influence of mind over mind? The instruments of this are private communication, the pulpit, & the press." How were the instructed to be selected? Who was qualified to do the choosing? The Saint-Simonians could offer no mode of institutionalizing the authority of the instructed that was consistent with their premises. All the same, Mill awarded them high marks for grasping the importance of the end.

The second distinctive contribution of the Saint-Simonians, Mill said, concerned the prominence they gave the connection between institutions and "a particular stage in the progress of the human mind." Certain institutions, considered independently of the conditions in which they developed and functioned, might seem woefully ill-suited for serving the cause of improvement. Only by taking such conditions into account can one properly estimate their value. When viewed in this light, these institutions might be found "the only means by which the human mind could have been brought forward to an ulterior stage of improvement" (the Catholic Church in the Early and High Middle Ages being a case in point). Essential for a candid assessment of the past, this perspective was also needed for an understanding of the present. The Saint-Simonian system encouraged a healthy "eclecticism" and "comprehensive liberality," a spirit of inquiry that labored against the baneful tendency to search for error in men's opinions when the object should be to open their eyes to that portion of the truth they had not yet perceived. (When discussing Saint-Simonism in this letter, Mill made no specific mention of Comte's work.) "The great instrument of improvement in men, is to supply them with the other half of the truth, one side of which only they have ever seen: to turn round to them the white side of the shield, of which they seeing only the black side, have cut other men's throats & risked their own to prove that the shield is black." A further virtue of the Saint-Simonian system, one directly associated with its awareness of the relation between institutions and stages of development, pertained to "the investigation of political truths." The Saint-Simonians rightly acknowledged the need "to ascertain what is the state into which, in the natural order of the advancement of civilisation the nation in question will next come; in order that it may be the grand object of our endeavours, to facilitate the transition to this state." The practical calculations and conclusions issuing from this method and purpose revealed that the way forward was usually anything but straight. Mill observed: "it will often follow, that we must uphold or even establish institutions which are liable to produce great evils, evils which in other states of society might be without alloy; provided that these institutions have, at the same time, a tendency to counteract other mischievous tendencies, which happen to be more prevailing or more to be apprehended in that age." For Mill, however, it was the state of the "nation in question" rather than the state of the human mind at large that had to be examined, and he reiterated his belief "that different nations, indeed different minds, may & do advance to improvement by different roads; that nations, & men, nearly in an equally advanced stage of civilization, may yet be very different in character, & that changes may take place in a man or a nation, which are neither steps forward or backward, but steps to one side."[74] (In these years Mill might have thought that a few of his own steps were to one side, or the other.)

Mill's further reading of *Le Prodecteur*, coupled with additional letters from d'Eichtal, had persuaded him that Saint-Simonian doctrine was still evolving, and that some of his initial objections had not been well-founded. Yet he discouraged his friend from holding out the prospect of a conversion that would give rise to Mill's active participation in the movement. Even if he became "convinced that the whole body of your doctrine is true," he still would not become a propagator of the creed. Mill explained why. "It appears to me utterly hopeless and chimerical to suppose that the regeneration of mankind can ever be wrought by means of working on their opinion." The societies of England and France would have to advance far beyond their current states before the teachings of Saint-Simonism, or any other comprehensive system of progressive thought, could obtain the results d'Eichtal and his colleagues looked for. Mill contended that the "adoption of St. Simonism, if that doctrine be true, will be the result and effect of a high state of moral and intellectual culture previously received." To present it to minds wholly unfit to fathom or practice its fundamental tenets would be futile.[75]

For England, certainly, Mill recommended a very different approach, one always mindful of the inability of his countrymen to understand or act upon any large lump of coherent doctrine. "Englishmen habitually distrust the most obvious truths, if the person who advances them is suspected of having any general views." How does a person holding such views manage to reach the audience he sought to influence? By homing in on discrete practical points, which, when pleasingly presented, could cumulatively help "form their habits of thought." One must first secure a "high reputation with them for knowledge of facts, & skill and judgment in the appreciation of details"; when this had been achieved one could "then venture on enlarged views; but even then, very cautiously and guardedly."[76] To promulgate a new doctrine, and then embark on a concerted drive to propagate its cardinal points, would accomplish nothing.

Mill assured d'Eichtal that his refusal to enlist in the Saint-Simonian cause did not mean that his loyalties lay elsewhere. This refusal rather signified Mill's opposition to putting himself "in the situation of an advocate for or against a cause." Advocacy was inseparable from controversy, and Mill said he wanted no part of "argumentation and debate." The habit of mind he wanted to cultivate required constructive engagement with a wide range of ideas. Far more could be gained by the gathering of truths than by the annihilation of errors. "I am averse to any mode of eradicating error, but by establishing and inculcating (when that is practicable) the opposite truth; a truth of some kind inconsistent with that moral or intellectual state of mind from which the errors arise. It is only thus that we can at once maintain the good that already exists, and produce more."[77] Mill's unwillingness to consider taking up the cause of Saint-Simonism was in this way presented as the particular expression of a general attitude.

Mill was here talking to himself as much as he was addressing d'Eichtal. This series of letters reveals his arduous struggle to adhere to the dialectic outlined above. Mill understood that he had a potent affinity with the critical spirit that he ascribed to the eighteenth century. His training had run with the grain of his mental powers and given him a formidable arsenal with which to engage in controversy. That he relished the use of this arsenal was evident not only in the articles he wrote during the mid-1820s but in his London Debating Society speeches of the late 1820s. In a debate of spring 1829 Mill and Sterling clashed fiercely when the latter inveighed against the immorality of Bentham and those who upheld his views within the London Debating Society. Mill reacted in a savage rejoinder that personally wounded Sterling.[78]

Sterling then left the debating society, and Mill worried that his attack on Sterling might have made a casualty of himself.[79] A fortnight after their verbal combat, he wrote to Sterling.

> I was unwilling that you should leave the London Debating Society without my telling you how much I should regret that circumstance if it were to deprive me of the chance not only of retaining such portion as I already possess, but of acquiring a still greater portion of your intimacy—which I value highly for this reason among others, that it appears to me peculiarly adapted to the wants of my own mind; since I know no person who possesses more, of what I have not, than yourself, nor is this inconsistent with my believing you to be deficient in some of the very things which I have.[80]

In part, Mill was determined to cultivate his friendship with Sterling for reasons not unlike those that moved him to assimilate "the reaction against the eighteenth century." Prone, in these years, to take stock of "the wants" of his "own mind," he acted to answer one of those wants. In another particular expression of the general attitude he had imparted to d'Eichtal, Mill himself withdrew from the London Debating Society at the end of 1829.

★ ★ ★

"I . . . am persuaded that discussion, *as* discussion, seldom did any good."[81] So John Mill had told d'Eichtal. Not much discussion passed between Mill and his father in the late 1820s; still less "argumentation and debate." From his "father's tone of thought and feeling, I now felt myself at a great distance," he notes in his *Autobiography*.[82] Neither the motive nor the content of John Mill's program of internal culture could elicit James Mill's understanding or support. Wordsworth, Coleridge, Sterling, the Saint-Simonians—such men had nothing to say to James Mill, and he had nothing to say to them. For his wife and children, not being spoken to by James Mill was perhaps a source of relief. Bain's generally sympathetic biography says of him: "In his advancing years, as often happens, he courted the affection of the younger children, but their love to him was never wholly unmingled with fear, for, even in his most amiable moods, he was not to be trifled with. His entering the room where the family was assembled was observed by strangers to operate as an immediate damper."[83] His eldest son could not muster the courage for a candid reckoning of the differences that had grown up between them. "On those matters of opinion on which we differed, we talked little."[84] A visitor to the Mill household in these years observed that John's manner with his father "was deferential, never venturing to controvert him in argument nor taking a prominent part in the conversation in his presence."[85] No reckoning, in any case, could restore an emotional confidence and trust that had never existed. "I expected no good, but only pain to both of us, from discussing our differences"; yet the younger Mill wanted readers of his *Autobiography* to know that he did not stand mute when his father "gave utterance to some opinion or feeling repugnant to mine, in a manner which would have made it disingenuousness on my part to remain silent."[86] Maintaining a modicum of self-respect demanded no less, and required no more. Whatever we make of this faint claim of assertiveness, James Mill no doubt sensed that a wall now separated his eldest son from him. He could not know why it had sprung up, or of what materials it was made.

The estrangement had more to do with feeling than with thought. In the region of elemental emotions John Mill found his father odious; the younger Mill's "opinions" on most political and philosophical subjects nonetheless had a lot in common with those of the elder. Mill's *Autobiography*, when discussing his intellectual development in these years, hints that he was striking out on a path of his own making, entering terrain his father had not trod. He sums up key changes in his thought, and attributes them to the new influences he had absorbed during the second half of the 1820s, influences he associated with "Continental" thought and "the reaction of the nineteenth century against the eighteenth."[87] He embraced the idea that "all questions of political institutions are relative, not absolute, and that different stages of human progress not only *will* have, but *ought* to have, different institutions"; another was that "any general theory or philosophy of politics supposes a previous theory of human progress; and this is the same thing with a philosophy of history."[88] Were such ideas disputed by James Mill? In his *History of British India* he stated: "No scheme of government can happily conduce to the ends of government, unless it is adapted to the state of the people for whose use it is intended."[89] Was James Mill's magnum opus innocent of a "theory of human progress"? In a letter written to his great friend Ricardo just months before the publication of *The History of British India*, he observed: "The subject afforded an opportunity of laying open the principles and laws of the social order in almost all its more remarkable states, from the most rude to the most perfect with which we are yet acquainted."[90] Did the author of the following passage, found in James Mill's *History*, suppose that his work had no "philosophy of history" of the sort alluded to by John Mill in his *Autobiography*?

> It is not easy to describe the characteristics of the different stages of social progress. It is not from one feature, or from two, that a just conclusion can be drawn. In these it sometimes happens that nations resemble which are placed at stages considerably remote. It is from a joint view of all the great circumstances taken together, that their progress can be ascertained; and it is from an accurate comparison, grounded on these general views, that a scale of civilization can be formed, in which the relative position of nations may be accurately marked.[91]

The idea of sequential stages in the historical development of civilization had a significant presence in the work of Adam Smith, Adam Ferguson, and John Millar, central figures in the Scottish Enlightenment. In many respects a child of this Enlightenment, James Mill readily accepted the idea, to which he early exposed his eldest son. By the time he was seven years of age, John Mill had already read Millar's *An Historical View of the English Government*, a book "highly valued" by his father.[92] When John Mill examined "the spirit of the age" in 1831, he essentially grafted the Saint-Simonian notion of alternating "organic" and "critical" periods onto a preexisting Scottish scheme of historical understanding.

Like the Saint-Simonians, Austin, and Coleridge, James Mill attached great weight to the influence of the instructed. He held that "In every society there are superior spirits, capable of seizing the best ideas of their times, and if they are not opposed by circumstances, of accelerating the progress of the community to which they belong."[93] James Mill adopted as his own the mission of advancing the progress of the community to which he belonged, and tried to work with others of like mind to bring this about. Moreover, he had educated his eldest son so as to prepare him for a leading part in the political and moral reformation of English society. This son would certainly have granted his father a place among those "capable of producing a beneficial effect on their age & country as teachers of the knowledge which fits people to perform their duties & exercise their rights, and as exhorters to the right performance & exercise of them."[94] Coleridge, Austin, and the Saint-Simonians, whatever the distinctive cast each gave the fundamental question of intellectual authority and its social influence, adhered to the basic premise asserted by James Mill concerning the vital importance of "superior spirits."

James Mill's most withering critic in the late 1820s was not someone John Mill would have linked with "the reaction of the nineteenth century against the eighteenth." In the mid-1820s James Mill brought out a volume consisting of essays he had originally written for the *Encyclopædia Britannica*.[95] Standing first was his *Essay on Government*, initially published in 1820. Received by John Mill and his fellow Benthamites "as a masterpiece of political wisdom,"[96] the *Essay* was seen by its author and his political disciples as "a concise and clear exposition of the Elements of Political Knowledge."[97] How to obtain the best security for beneficial legislation was the chief problem examined in the *Essay*. James Mill argued, in typically Benthamite fashion, that the answer could be deduced from a universal attribute of human nature: the pursuit of individual self-interest. Power in the hands of a minority would inevitably be used to profit those composing the privileged group, who would oppress the majority to satisfy their own selfish ends. It could not be otherwise in a world where "every man, who has not all the objects of his desire, has inducement to take them from any man who is weaker than himself."[98] To prevent such abuses, and secure legislation beneficial to the community as a whole, a system of representation had to be devised that gave effectual political expression to the aggregate of individual interests present within the community. The members of the representative body must be chosen in such a way as to identify the interests of those responsible for making public policy with the interests of the community at large. An extensive suffrage and frequent elections were the instruments necessary for accomplishing this essential end. James Mill conceded that a suffrage less than universal could attain this goal. Women could be excluded, their interests being subsumed by those of their fathers or husbands; so too could men under the age of forty, on the grounds that their interests substantially overlapped with those of similarly situated men above that age.[99] Frequent elections were indispensable to binding the interests of the representatives to those of the electors. The premises and conclusions of the *Essay on Government* came under fierce assault in 1829, when T.B. Macaulay moved to detonate James Mill's "Utilitarian Logic" in the pages of the *Edinburgh Review*. The ensuing debris would cause John Mill to question the adequacy of his father's mode of political reasoning.

The leading luminary among the younger set of Edinburgh Reviewers, Macaulay had not yet reached his thirtieth birthday when he set his sights on the redoubtable James Mill. Although not personally well-acquainted with the elder Mill, Macaulay did know some of the junior members of the Utilitarian circle. Like many others, he had felt

Charles Austin's influence at Cambridge, both men declaiming brilliantly in the debates of the Cambridge Union. Macaulay's sister Hannah noted that Austin had successfully steered her brother toward more liberal views during their years at university.[100] In the late 1820s Macaulay was one of several recent Cambridge graduates recruited by John Mill for the fortnightly meetings of the London Debating Society. Roebuck and Mill were frequent speakers, and Macaulay's exposure to their political views amplified his awareness of James Mill's influence. Macaulay's extraordinary intellectual and oratorical abilities had been evident in childhood, not least to himself, and his supreme confidence in his powers of thought and expression seemed apt to those who witnessed his performances at the Cambridge Union.[101] Physically unprepossessing, mentally he was a torrential force. Relative youth proved no bar to Macaulay's joining issue with James Mill in 1829.

In fact Macaulay did not wait until 1829 to assail the politics of James Mill and his acolytes. Two years before, in an essay for the *Edinburgh Review*, he had voiced his keen anxieties regarding the presence, within the "middling orders," of "a Republican sect, as audacious, as paradoxical, as little inclined to respect antiquity, as enthusiastically attached to its ends, as unscrupulous in the choice of its means, as the French Jacobins themselves,—but far superior to the French Jacobins in acuteness and information—in caution, in patience, in resolution." The members of this sect, he continued, were contemptuous of the "merely ornamental," and had an antipathy to the fine arts, graceful literature, and the "sentiments of chivalry." Their intellectual presumption had "made them arrogant, intolerant, and impatient of superiority."[102] In this article Macaulay did not put a name to the sect he had in mind or identify its leaders; any doubt on this score, however, was removed by his 1829 article on James Mill, which attributed precisely these qualities to the Utilitarians, and in much the same language.[103]

Macaulay's alarm over the conduct of the Utilitarians stemmed in part from his sense of the social and political order they sought to radicalize. He saw this order as perilously unstable. The history of England since Waterloo, he declared in the essay of 1827, "is almost entirely made up of the struggles of the lower orders against the government, and of the efforts of the government to keep them down."[104] The years after the return of "peace" had been tumultuous. Appalling hardship had given rise to widespread disturbances that in turn prompted government crackdowns. In 1817 Lord Liverpool's administration introduced, and Parliament enacted, temporary legislation suspending habeas corpus. Further Coercion Acts, including measures authorizing magistrates in disturbed districts to search for arms and to forbid the holding of mass meetings, were passed in 1819 in the wake of the so-called Peterloo Massacre. (On 16 August an attempt by local yeomanry to break up a huge reform demonstration at St. Peter's Field outside Manchester had left some dozen people dead and hundreds more seriously injured.) In 1820 came the mass demonstrations in support of the new king's long-estranged wife Caroline, who had returned from the Continent to lay claim to her place as queen, a claim George IV could not abide. The admonition Macaulay gleaned from this last episode underscored his concerns about the trouble-making potential of the Utilitarians:

> On that occasion, the majority of the middling orders joined with the mob. The effect of the union was irresistible. The Ministers and the Parliament stood aghast; the bill of pains and penalties [against Caroline] was dropped; and a convulsion, which seemed inevitable, was averted. But the events of that year ought to impress one lesson on the mind of every public man,—that an alliance between the disaffected multitude and a large portion of the middling orders, is one with which no government can venture to cope, without imminent danger to the constitution.[105]

Macaulay viewed the Utilitarians as the vanguard of such an alliance, the formation of which he hoped to see thwarted.

Only prudent and judicious reform, the sort favored by the Whigs, could preserve the balance of the constitution. So Macaulay believed. The events that occasioned his essay of 1827 flowed from the recent stroke that had disabled Lord Liverpool, the Tory premier since 1812. For some years Liverpool had skillfully—and not without great strain—held together a ministry hampered by deep personal distrust and by absence of consensus on the issue of whether or not Catholics should be allowed to sit in the House of Commons. George Canning, Liverpool's foreign secretary and leader of the House of Commons, headed a wing of the cabinet that supported Catholic emancipation. Also in the government were enemies of both Canning and the Catholic cause. Liverpool, generally a friend to Canning, lined up with Canning's opponents on the Catholic question, and through his personal influence and authority he had managed to keep the government afloat. With Liverpool incapacitated, Canning succeeded to the premiership. Some of his former colleagues refused to serve under him, and Canning therefore had to seek Whig participation in his government. In a move endorsed by Macaulay and the *Edinburgh Review*, a handful of Whigs accepted office.[106] The only alternative to such a coalition, Macaulay argued, was a reactionary Tory administration, "ignorant and tyrannical." Should this come to pass, the radical principles of the Utilitarians "would spread as rapidly as those of the Puritans formerly spread, in spite of their offensive peculiarities. The public, disgusted with the blind adherence of its rulers to ancient abuses, would be reconciled to the most startling novelties."[107]

It may seem odd that Macaulay should speak of the Utilitarians and the French Jacobins in the same breath, and also suggest that the main difference between them lay in the greater cunning of the former. For Jacobinism, of any stripe, James Mill had as little use as Macaulay. Although he was ready to exploit the fear of violent revolution within the governing classes to further the cause of reform, the coming of such an event would have made him shudder.[108] James Mill saw thoroughgoing legal and political reform, mingled with mass education, as the pivotal forces for bringing about significant human improvement. The less thoroughgoing approach of the Whigs did not prevent him from doing his utmost to assist the passage of bills for parliamentary reform introduced by Lord Grey's government in 1831 and 1832. Macaulay would be one of the most eloquent defenders of these measures, and the argument he then made for the enfranchisement of the middle classes sat comfortably with much of the case put forward by James Mill in his *Essay on Government*. Henry Brougham, the Whig politician and fellow Edinburgh Reviewer with whom Macaulay most closely identified in the late 1820s, worked in harness with his friend James Mill in support of the British and Foreign School Society, the infant school movement, the Society for the Diffusion of Useful Knowledge, and the University of London.[109] To the creation of that nonsectarian university Macaulay gave a hearty and celebratory welcome in the pages of the *Edinburgh Review*.[110] In the years before James Mill made his name as the author of *The History of British India*, his occasional articles for the *Edinburgh Review* brought in some badly needed income. Macvey Napier, the man who originally commissioned James Mill's essays for the *Encyclopædia Britannica*, later succeeded Francis Jeffrey as editor of the *Edinburgh Review*.

These encyclopædia essays, to be sure, were solicited before the launching in 1824 of the *Westminster Review*, whose earliest numbers featured scorching attacks on the *Edinburgh Review*. The first of these was written by James Mill, the second by his eldest son under the father's supervision.[111] The thesis they proposed rejected as mythical the notion that the British Constitution struck a balance among the monarchy, aristocracy, and people; in practice, governmental power resided in the aristocracy alone. Parliament, the law, the church—all were political instruments designed to perpetuate aristocratic domination. The two parties, Tory and Whig, were simply two factions of the oligarchy, each of them determined to shield and succor the sinister interests of the Few against the legitimate interests of the Many. Whereas the Tories scarcely bothered to conceal their blatant bias, the Whigs tried to present themselves as the friends of the people and the balanced constitution. This ambition led them to ride a political "see-saw," one that on critical issues routinely tilted in favor of the aristocracy. To demonstrate the validity of this contention, the Mills canvassed a wide range of essays printed in the *Edinburgh Review*, the chief periodical organ of the Whigs. They claimed to show that the conduct of the *Edinburgh*, from its inception to the present, amply exhibited the working of the "see-saw" and the unmistakable refusal of the Whigs to come to grips with the real abuses of aristocratic government. "What can be more immoral than the see-saw? A practice which is, throughout, a mere sacrifice of truth to convenience: a practice which habituates its votaries to play fast and loose with opinions—to lay down one, and take up another, with every change of audience?"[112] The readiness of leading Whigs to join forces with Canning in 1827, and of the *Edinburgh Review* to justify this course of action, conformed well enough to this interpretive scheme. Yet the Whigs and the *Edinburgh* understandably deemed these charges unfair and offensive. Their purpose, as they conceived it, was to defend constitutional liberty and the cause of prudent reform against the toxic threats of both despotism and democracy. The Utilitarians should not have been surprised when the *Edinburgh* retaliated. What they could not have foreseen was the potency of Macaulay's assault on the foundations of their political reasoning.

According to Macaulay, the author of the *Essay on Government* showed no interest in testing his sweeping premises against the experience of governments that had actually existed. James Mill began with a set of assumptions about certain predilections inherent in human nature, "and from these premises the whole science of Politics is synthetically deduced!"[113] Such a method might assist James Mill in reaching the conclusions he wished to arrive at, but neither the assumptions nor the conclusions did anything to explain the variegated motives from which conduct in the real world springs. The self-interest principle, treated by Mill as a law regulating political phenomena, had neither explanatory nor predictive value. It signified only that "men, if they can, will do as they choose." The actions of a man reveal something about what he conceives his interest to be; what we may think of as his interest cannot be relied upon to tell us what his actions will be. "One man goes without a dinner, that he may add a shilling to a hundred thousand pounds: another runs in debt to give balls and masquerades. One man cuts his father's throat to get possession of his old clothes: another hazards his own life to save that of an enemy. One man volunteers on a forlorn hope: another is drummed out of a regiment for cowardice."[114] Furthermore, James Mill's idea that the interests of women could be safely entrusted to their fathers or husbands could not be squared with either a large portion of human experience or with the dictum that the desire to exploit others in the service of our own pleasure was embedded in human nature. On James Mill's own premises, would not a democratic suffrage produce a plundering of the rich by the poor? And on what grounds could it be assumed that power lodged in the hands of the middle classes would be used for less selfish ends than power lodged in the hands of any other group? James Mill's speculative method could not yield "just conclusions," which could only be reached by induction, the method used "in every experimental science" to augment "the power and knowledge of our species." Nothing of value could emerge from an attempt to build a science of politics on so-called laws of human nature. Conversely, much good could come from "observing the present state of the world"; "assiduously studying the history of past ages"; "sifting the evidence of facts"; "carefully combining and contrasting those which are authentic"; "generalizing with judgment and diffidence"; "perpetually bringing the theory which we have constructed to the test of new facts"; "correcting, or altogether abandoning it, according as those new facts prove to be partially or fundamentally sound."[115]

In his *Autobiography* J.S. Mill discusses the impact of Macaulay's attack on his own view of his father's manner of political reasoning. The theory of government propounded by Bentham and James Mill had of course come under fire before Macaulay's essay appeared; John Mill noted that his own part in defending this theory, together with his recent discovery of quite different "schools of political thinking," had already disposed him to think Utilitarian doctrine imperfect. Without concluding that James Mill's theory of politics was unsound, he had decided that corrections were needed in "applying the theory to practice." Macaulay's sortie, however, struck at the foundations of the theory, and several of his criticisms hit the mark: James Mill's premises, his son acknowledged, "were really too narrow, and included but a small number of the general truths, on which, in politics, the important consequences depend." Good government could not be achieved simply through aligning the interest of the representative body with that of the community, and electoral provisions in themselves could not ensure such an identity of interest.[116]

John Mill considered his father's response to Macaulay's animadversions far from satisfactory. James Mill should have answered that his critic had mistaken "an argument for parliamentary reform" for "a scientific treatise on politics." (Surely James Mill saw his *Essay* as both these things, and presented his general reform proposals as the practical corollaries of the science whose rudimentary principles he had stated. He would have scoffed at the idea of giving such an answer as his son would have had him give. Presumably John Mill did not offer his father advice on this matter, or any other.) How did his father react? "He treated Macaulay's argument as simply irrational; an attack upon the reasoning faculty; an example of the saying of Hobbes, that when reason is against a man, a man will be against reason." This retort unsettled John Mill, causing him to "think that there was really something more fundamentally erroneous in my father's conception of philosophical Method, as applicable to politics, than I had hitherto supposed there was."[117]

Macaulay's attack came at a time when John Mill was inclined to probe for weaknesses in his father's way of thinking. Weaknesses he found, but they hardly persuaded him that Macaulay had got right what his father had got wrong. James Mill's premises being deficient in scope and plenitude did not mean that one could contrive to do without premises altogether. Macaulay's crude brand of induction, which called for an exclusively empirical treatment of political phenomena, was fundamentally misconceived. The plurality and complexity of the causes and effects characteristic of such phenomena rendered the experimental method proposed by Macaulay of no use whatsoever. Indeed, in his *System of Logic* John Mill scornfully rejected the mode of proceeding urged by Macaulay (he no doubt had the latter in mind when expressing this scorn): "The vulgar notion, that the safe methods on political subjects are those of Baconian induction—that the true guide is not general reasoning, but specific experience—will one day be quoted as among the most unequivocal marks of a low state of the speculative faculties in any age in which it is accredited."[118] Although the particular form of "general reasoning" applied by James Mill to political affairs was ill-adapted to the purpose, the science of politics had to be deductive. In his 1835 essay "Rationale of Representation," John Mill asked: "If principles of politics cannot be founded, as Burke says, 'on the nature of man,' on what can they be founded? On history? But is there a single fact of history which can be interpreted but by means of principles drawn from human nature?"[119]

Of course the epistemology of the Mills, father and son, posited that all knowledge, including knowledge of human nature, derived from "experience." The formulation of sound abstract principles depended on "the accumulated results of experience."[120] The power of human observation, steadily and rigorously applied, gave rise to serviceable general theories. The younger Mill would argue that the science of politics must rest on approximate generalizations, a category of inductive truths from which probable inferences could be drawn. Unlike the universal truths of the physical sciences, these approximate generalizations could not be relied upon to account for every individual instance of a given phenomenon. The science of politics, however, dealt with the actions of mass communities rather than with those of single individuals. The person seeking to understand and influence such actions could "get on well enough with approximate generalizations on human nature, since what is true approximately of all individuals is true absolutely of all masses."[121] James Mill deduced too much from too few approximate generalizations, an error less serious than that made by Macaulay, who discarded generalization entirely in the mistaken belief that empiricism alone could furnish a safe guide to practice.

It should be noted that John Mill's caveats regarding his father's theoretical stance produced no separation in the realm of practical politics. Macaulay and John Mill both understood that James Mill had one eye, maybe two, on English politics and society when he wrote the *Essay on Government*. Relative to England, John Mill was persuaded that his father's suppositions and policy recommendations were authoritative. He states in the *Autobiography* that on practical political questions he and his father "were almost always in strong agreement."[122] John Mill remained "a radical and democrat, for Europe, and especially for England."[123] The truths gathered by Mill in these years did not undermine the fundamentals of his practical political creed, a creed delivered to him by his father.

★ ★ ★

As the 1820s drew to a close, John Mill, for all his advantages, had somber thoughts regarding his prospects for personal happiness. Writing to Sterling in April 1829, he alluded to "the comparative loneliness of my probable future lot." This sense of loneliness, he explained, did not arise from "misanthropy" or a sense of social isolation. "At present I believe that my sympathies with society, which were never strong, are, on the whole, stronger than they ever were." The lack he felt so keenly at this time, one he feared would not be made good, originated in his conviction of superiority. "There is now no human being (with whom I can associate on terms of equality) who acknowledges a common object with me, or with whom I can cooperate even in any practical undertaking without the feeling, that I am only using a man whose purposes are different, as an instrument for the furtherance of my own."[124]

Eyton Tooke, a dear friend of John Mill's since the early 1820s, apparently was not someone with whom Mill could "associate on terms of equality." In his letter to Sterling, Mill implied that he had outgrown Benthamite doctrine and left behind the coterie of young men, Tooke included, whose comradeship had stemmed from belief in a common cause. Near the end of January 1830 Eyton Tooke committed suicide. Mill's reaction, as evinced in a letter to d'Eichtal written a fortnight after "the horrible event," was that of someone whose feelings were difficult to distinguish from the thoughts he had about his feelings.[125] He acknowledged that he had been "deprived ... of one whom I had counted upon as a friend and companion through my whole life." Even so, he assured d'Eichtal that the blow, although acutely felt, had not made him utterly desolate. "Many who knew him and loved him less than I did, have felt the immediate shock much more forcibly." Mill went on to explain that his grief made up in "durability" for what it lacked in "intensity" (an odd observation given the relative proximity of the terrible deed?). "I feel it most in enervation, & almost extinction, for the present, of all my activity, and all my concern for mankind or for my duties. It seems to me as if I had never cared for any one but him, and had never laboured but for his sympathy and approbation." Yet Mill apprehended the transitory nature of this disposition. "I know this effect of it will not last. The more affectionate I cherish his memory, the more ardently shall I pursue those great objects in which he took so deep an interest." Mill then introduced the theme he had brought forward in his letter to Sterling: the significance he attached to the experience of fellowship in a common cause. "I should care little for life or for mankind, but for the thought that there are among them a few men like him [Tooke], & that all have in them the capacity of becoming at least an approximation of what he was. There are yet two or three living, but for whom I should no longer value existence." He conceded that "scarcely any" of these two or three "were equal to Eyton and none superior to him in purity and singleness of mind combined with warmth and kindliness of affection, yet I cannot entirely droop or relax in my exertions while they survive, and remain unchanged towards me, and progressively improving in the developement of their intellectual and moral capabilities."[126] Mill's letter to d'Eichtal underscored the inestimable importance to him of strong friendship and the great worth he had placed upon his bond with Eyton Tooke. It also intimated that the grave loss he had suffered would not imperil his own existence.

The tribute paid Eyton Tooke in the *Autobiography*—"a young man of singular worth both moral and intellectual"—fittingly summed up Mill's judgment of Tooke's moral qualities, without revealing his true opinion of Tooke's intellectual abilities.[127] Mill's letter to d'Eichtal had cited Tooke's "purity and singleness of mind combined with warmth and kindliness of affection"—laudable traits, no doubt, but not the stuff of intellectual distinction. Less than a month after writing to d'Eichtal, Mill raised the matter of Eyton Tooke's death in a letter to the eminent French political economist J.B. Say, whose guest the young John Mill had been at the start and end of his stay in France in 1820–21. After expressing his condolences to Say upon the recent death of Madame Say, John Mill feelingly noted his own bereavement.

> I have myself suffered a most grievous and unexpected loss, by the death of my poor friend Eyton Tooke, who was well-known to you as one of the most excellent and promising of all his contemporaries, and who would have been a blessing to his country and to his kind. The loss of such a man will be felt in a thousand ways by persons who never knew him nor were aware what things were to be expected from him if he had lived to pursue the career of self-improvement and philanthropic exertion which he had entered upon; and how admirable a moral influence he would have exercised on all with whom he came in contact, by the unrivalled purity and rectitude of his purposes, combined with the largest and most comprehensive liberality and philanthropy.[128]

Had Mill thought Eyton Tooke's intellectual power proportionate to his moral standing, he would have said so in his letter to Say. Tooke's limitations were explicitly acknowledged in a letter Mill wrote John Pringle Nichol in December 1834. Nichol, a young astronomer with a vigorous interest in political economy, was evidently on the mend after a bout of serious illness and depression. Mill noted: "I once lost a most valued friend, one of the most valued I ever had—though not to be compared with you in intellect—in consequence of a similar disease—the eldest son of Tooke, the political economist."[129] Tooke's uncommon virtue had not been enough to put him on the elevated plane occupied by John Mill.

Ultimately John Mill would choose to believe that one person in his life ranked not only as his equal but as his superior. He met this person in the same year Eyton Tooke died. Whether or not Harriet Taylor added to or subtracted from the advantages John Mill enjoyed in the late 1820s is a question he would have thought absurd when he came to write his *Autobiography*. For some time after 1830, however, the question remained open.

CHAPTER THREE

Mill and Harriet Taylor: The Early Years

When John Mill entered the orbit of Harriet Taylor he was twenty-four years of age, and she seventeen months his junior. He was a young man whose extraordinary intellectual gifts had already been acknowledged in certain circles; she was part of a circle whose most prominent member, William Johnson Fox, appreciated the worth of John Mill's gifts and the varied use to which they could be put. A charismatic Unitarian preacher and impassioned political radical, Fox was pastor of the South Place congregation in Finsbury. He also edited the *Monthly Repository*, the periodical organ of the British and Foreign Unitarian Association. Unitarianism and Utilitarianism, movements integral to the English Enlightenment, shared a powerful commitment to the causes of civil equality, religious liberty, free inquiry, educational reform, and general human improvement. Unitarian and Benthamite circles conspicuously overlapped. Both John Bowring and Henry Southern, coeditors of Bentham's *Westminster Review*, were Unitarians. When the first issue of the *Westminster Review* appeared in January 1824, the author of its opening article was none other than W.J. Fox. When Jeremy Bentham died in June 1832, his funeral oration was delivered by Dr. Thomas Southwood Smith, quondam Unitarian minister and a man still closely associated with Fox.[1] A common Unitarian upbringing underlay the marriage of John and Sarah Austin. The world of respectable London radicalism familiar to the young John Mill incorporated prominent strains of Unitarianism and Utilitarianism. There was nothing especially odd in his being invited to dine at the home of John Taylor and his young wife Harriet, two of W.J. Fox's most valued congregants.

A prosperous wholesale druggist in a well-established firm based in the City of London, John Taylor was a man of notable liberality, a stalwart supporter of W.J. Fox and the South Place Chapel, and a generous friend to political refugees from the Continent who had made their way to London. He had wed Harriet Hardy in 1826, when he was nearly thirty years of age and she just eighteen. She had grown up in Walworth, south London, where her father practiced as a surgeon and "man-midwife." Her mother, née Harriet Hurst, came from a family that proudly claimed descent from Cavalier stock. Tense relations with her parents and many of her siblings (she had five brothers and one sister) may have predisposed her to accept a matrimonial proposal from a man of abundant good nature and ample means. It is not hard to account for John Taylor's interest in making such a proposal. A daughter of W.J. Fox vividly recalled the aura cast by Harriet Taylor when the latter was in her mid-twenties (a portrait, evidently painted when Harriet Taylor was approximately this age, corroborates the word-picture).

> Mrs. Taylor at this date, when she was, perhaps about five and twenty years of age, was possessed of a beauty and grace quite unique of their kind. Tall and slight, with a slightly drooping figure, the movements of undulating grace. A small head, a swan-like throat, and a complexion like a pearl. Large dark eyes, not soft or sleepy, but with a look of quiet command in them. A low sweet voice with very distinct utterance emphasized the effect of her engrossing personality.[2]

The first years of the marriage seem to have brought happiness to both parties. Their first child, a son (Herbert), was born in 1827. A letter Harriet wrote to her husband in July 1828, when she was visiting family on the Isle of Wight, was all that a loving husband could hope for. "Do not imagine my dearest that I ever doubted that your wish for our uniting was as great as mine. I knew that my dear husband loves me, as I have loved him, with my whole heart . . . I think from my present feelings that I shall never again consent to our parting."[3] In fact, she did consent to another parting, and fairly soon after. In the summer of 1829 Harriet Taylor, pregnant for the second time, was once more on the Isle of Wight, her husband having remained in London.[4]

Something about her marriage had begun to trouble Harriet Taylor, and she sought the counsel of Fox, her spiritual guide and a man whose radicalism included advocacy of equal rights for women. According to Alexander Bain, "Mrs. Taylor made Fox her confidant as to her want of sympathy from her husband... and Fox suggested her becoming acquainted with Mill."[5] Bain implied that his source was Harriet Martineau, who in the late 1820s and early 1830s wrote extensively for Fox's *Monthly Repository*. She too was a guest at the dinner to which Mill and his friends Roebuck and Graham were invited, as was Fox himself. Martineau, an inveterate gossip, apparently had many more things to say regarding the circumstances, but Bain concluded that there could be "no good in repeating them."[6]

What Bain understood the vague phrase "want of sympathy from her husband" to mean is not clear. No one has ever suggested that the cause of the problem lay in John Taylor's want of feeling for his wife, or that the husband rather than the wife might have approached Fox about the state of the marriage. Harriet Taylor had an elevated opinion of herself, one she found frequently confirmed by the admiration she easily elicited. She wanted for herself a creative and purposeful life, one befitting a person of exceptional qualities. Her notion of what such a life should involve had no doubt changed in the years since she had wed John Taylor and become associated with Fox's coterie, which included a number of talented young women whose aspirations and activities transcended the domestic sphere. An exceptional woman, she believed, could find true companionship only with an exceptional man. Before the end of 1830 she had concluded that John Taylor, for all his worthy qualities, could not rise to such a standard. By then she was again with child. Three pregnancies in less than five years had put paid to what little sexual ardor she might once have felt for her husband. His physical enthusiasm for his beautiful young wife, on the other hand, may well have continued robust. Even had a degree of compatibility been created and maintained in this sphere, it would not have kept Harriet Taylor from becoming disgruntled. She had a keen appetite for ideas, poetry, art, and music, and her affiliation with the *Monthly Repository* circle heightened this appetite. In feeling and in thought (the former especially), Harriet Taylor needed to soar. John Taylor, whom Thomas Carlyle described as "an obtuse most joyous-natured man, the pink of social hospitality,"[7] could do many valuable things; soar he could not.

What did Fox think might follow from bringing Harriet Taylor together with John Mill? One thing he probably hoped for related to his own aspirations. Harriet Taylor's "engrossing personality" could help Fox snare Mill for the *Monthly Repository*, which Fox wanted to make less denominational and more literary and political. In serving himself, however, Fox had no trouble believing that he could also benefit others. John Mill and Harriet Taylor, he reckoned, would find one another's company enjoyable and stimulating, and the growth of a friendship between the two would boost Harriet Taylor's flagging spirits. Her marriage to John Taylor might be more easily endured if Mill's attentions nourished her mind and emotions. Fox presumably understood that risk also attended the dinner to which John Mill and his friends were invited. Its presence did not deter Fox, whose views regarding the "sanctity" of marriage were unorthodox (in 1833 he would use the pages of the *Monthly Repository* to argue against the indissolubility of marriage).[8] Indeed, Fox's ambiguous personal circumstances in the early 1830s could well have figured in the gossip later dispensed by Harriet Martineau.

Fox had married in 1820 (he was then in his mid-thirties). Two years later he experienced a total breakdown, and for a time he could not preach. If marital strain did not cause this collapse, it contributed to the severity of his illness.[9] Fox gradually recovered his health, but his personal affairs took on a further complication after the Foxes were introduced to Benjamin Flower—a radical printer, editor, and writer—and his two exceptionally gifted and exquisitely beautiful daughters, Eliza and Sarah. (Eliza would achieve a measure of distinction as a composer; Sarah as a poet, a writer of hymns, and an actress. They apparently inspired the creation of the Ibbotson sisters in Harriet Martineau's 1839 novel *Deerbrook*.) Benjamin Flower's wife had died in 1810, when Eliza was seven and Sarah five, and he took charge of their education. This education did not include much in the way of disciplined instruction; it did, however, allow for the unimpeded flow of his daughters' impressive natural endowments. Benjamin Flower had Unitarian connections, and a reputation as a fearless radical (in 1799 he had been sent to prison for six months for his alleged libel of an Anglican bishop). In 1820 he and his daughters moved to Dalston, near Hackney, and in the years that followed all three were drawn increasingly into the London milieu of radical Unitarianism. After 1823 they became notable members of the South Place Chapel circle. Benjamin Flower and W.J. Fox cultivated a warm friendship. When Flower died in 1829 his daughters joined the household of the unhappily married Fox, who had been named their trustee. Eliza Flower became many things to William Fox—editorial assistant, amanuensis, daily companion, and more. They loved one another, and the love was not that of father and daughter. Mrs. Fox, not without good reason, came to feel that Eliza Flower had usurped the place that rightfully belonged to her.

The daily personal, social, political, and intellectual world of W.J. Fox scarcely made room for the marginalized Mrs. Fox, the mother of his three children. Eliza Flower had a central place in that world. Harriet

Taylor had struck up a great friendship with Eliza Flower sometime before the dinner in 1830 that brought John Mill and his two friends to the Taylors' Finsbury residence in Christopher Street.[10] Whatever Fox knew of Harriet Taylor's marital situation, Eliza Flower also knew. Both would want to promote the happiness of their dear friend. To John Taylor's interests they probably did not give a lot of thought. Their circumstances, and their sympathies, would not lead them to worry unduly about the threat to those interests that John Mill's entry into the Taylor household might represent.

Mill, of course, did not see himself as threatening anyone's interests when he accepted the Taylors' invitation to dine. His connection with Fox at that juncture was more political than personal, and even the political connection had not been unusually close. There is no reason to think that Fox told him in advance that he was likely to find Mrs. Taylor's company more bracing than Mr. Taylor's. She was armed with intelligence supplied by Fox: John Mill was brilliant, cultivated, high-minded, and sympathetic. Provided he did not disappoint on first acquaintance, Mill's favor and admiration would be worth having. Harriet Taylor's power to captivate should not be underestimated. She possessed elegant beauty, an abundant self-confidence, a vigorous intellect, and a ready wit. Her very manner of being imparted a distinctive force. After meeting Harriet Taylor for the first time, in the summer of 1834, Thomas Carlyle enthused (in a letter to his brother John): "she is a living romance-heroine, of the clearest insight, of the royallest volition; very interesting, of questionable destiny, not above twenty-five: Jane [Carlyle's wife] is to go and pass a day with her soon . . . being greatly taken with her."[11] The young John Mill was ripe for an awakening of the heart, and Harriet Taylor had the means to arouse in him feelings such as he had never before known.[12]

Or had he? Mill, in the *Autobiography*, refers to the "small circle of friends" with whom Harriet Taylor was on familiar terms at the time they met. Of these friends, only one "was a person of genius, or of capacities of feeling or intellect kindred with her own."[13] Mill does not name this person ("long since deceased"), but the obvious candidate is Eliza Flower, Harriet Taylor's most cherished friend (before Mill himself came along). Mill "reviewed" several of Flower's musical compositions for the *Examiner* in the early 1830s. The following passage, from a review of her *Songs of the Seasons* (1832), typifies his lavish praise of the music and its composer.

> There is in every strain that which denotes it to be the work not merely of an accomplished musician, but of a mind penetrated with

the spirit of a true artist . . . The language which Miss Flower has chosen is music, and she speaks it like one to whom none of its dialects is unfamiliar, because none of the feelings to which it is appropriate, from the loftiest to the most tender, is a stranger to her. When to these qualities of the inmost nature, is added originality of melody, together with adequate scientific knowledge of harmonic principles, we have all which constitutes musical genius, in the highest and most exclusive sense of the term.[14]

Friendship played a part in moving Mill to puff the creations of Eliza Flower; he felt he could do this, however, without abandoning sincerity. Her artistic power, as he saw it, had the imprint of genius.

No contemporary documents point to a personal acquaintance between Mill and Eliza Flower before he came to know Harriet Taylor. In the 1890s Moncure Conway, whose ministry of South Place Chapel spanned most of the last thirty-five years of the nineteenth century, brought out a book commemorating the centenary of South Place. In its pages he spoke of Eliza Flower: "John Stuart Mill was supposed at one time an aspirant for Eliza's hand; but she was the spouse of her art, consecrated to its ideal."[15] Conway, born in Virginia in 1832, had never met Eliza Flower, who died of consumption in 1846. He revered Mill, with whom he became friendly during the last decade of Mill's life.[16] Even if Mill had at one time felt a romantic interest in Eliza Flower, it is most unlikely that he would have mentioned the matter to Moncure Conway. The story Conway reported in his history of South Place presumably existed as a sensational element in the congregation's oral tradition.

If credence be given this story, certain items in the established historical record can be used to contrive a different sort of narrative relative to the start of the friendship between Mill and Harriet Taylor. Eliza Flower's rejection of a marriage offer might have precipitated Mill's prophecy, confided to Sterling in April 1829, that his "probable future lot" would be one of "comparative loneliness." Fox's initiative in bringing Mill together with Harriet Taylor could have been influenced by his wish to take Mill's mind off Eliza. Mill's disposition to feel his heart swell when Harriet Taylor displayed an ardent interest in him could have been colored by his disappointment at the hands of Eliza Flower. An alternative chronology could place the dinner at the Taylors before a Millian approach to Eliza Flower. This gathering could then be seen as the occasion marking Mill's introduction to the South Place circle, in which the Flower sisters held a prominent presence. Not a lot is known about the first two years of Mill's association with Harriet Taylor, which might have developed into a serious relationship only after Mill had pursued Eliza Flower and been rebuffed. None of this, however, seems plausible. Had Mill proposed marriage to Eliza Flower before April 1829, his involvement with Fox's group would have had to begin well before anything in the documentary record indicates was the case. Had Fox wanted to turn Mill's thoughts away from Eliza Flower, would he not have tried to keep him at arm's length rather than arrange a dinner at the home of her intimate friend Harriet Taylor? Moreover, it is hard to imagine that Harriet Taylor could have been kept uninformed of a marriage proposal issued by Mill to Eliza. Would she have considered Mill worthy of her love had he declared his devotion to another? Not likely.

Moncure Conway's report has all the earmarks of the apocryphal. The lore of South Place made much of Eliza Flower's beauty and genius, and Conway would have heard a lot about her extraordinary qualities during his decades as its minister. Mill, dead some twenty years at the time Conway published his account, would have been remembered as a figure of towering eminence, and by then his admiration for Eliza Flower's person and art would have been bandied about by several generations of South Place communicants. The parties directly involved circa 1830—Mill, the Flower sisters, Fox, the Taylors—had long since left the scene, and were apt to become the stuff of legend in a congregation such as South Place. The idea that Mill had at one time proposed to Eliza Flower could readily take hold of the imagination, and nothing more than embellished hearsay would be needed for such a notion to gain currency. Conway accepted its truth; we should not.[17]

Thomas Carlyle never linked Mill romantically with Eliza Flower. Harriet Martineau, with whom the Carlyles developed a friendship during the second half of the 1830s, would not have neglected to tell him of such a connection had one existed. To suppose that one existed, and that Martineau had not learned of it, does scant justice to her talent for acquiring piquant intelligence. Of the initial meeting of John Mill and Harriet Taylor she had first-hand knowledge, and Carlyle adapted Martineau's description of the encounter for his own sardonic purposes directly following Mill's death in 1873. Walking with Charles Eliot Norton, the American man of letters, Carlyle gave this account:

> [John] Taylor was a verra respectable man, but his wife found him dull; she had dark, black, hard eyes, and an inquisitive nature, and was ponderin' on many questions that worried her, and could get no answers to them, and that Unitarian clergyman you've heard of, William Fox by name, told her at last that there was a young

philosopher of very remarkable quality, whom he thought just the man to deal with her case. And so Mill with great difficulty was brought to see her, and that man, who up to that time, had never looked a female creature, not even a cow, in the face, found himself opposite those great dark eyes, that were flashing unutterable things, while he was discoursing the utterable concernin' all sorts o' high topics.[18]

The imperatives of Carlylean caricature, coupled with the animosity he ultimately came to feel toward Harriet Taylor, moved Carlyle to emphasize Mill's innocence and Taylor's "great dark eyes . . . flashing unutterable things." Harriet Taylor must have done some "discoursing" of her own, and Mill no doubt liked what he heard. (If he liked what he saw even more than what he heard, he would not have admitted this to himself.) Unlike Carlyle, Roebuck was present at the dinner given by the Taylors. He failed to notice anything unusual in "what passed that evening, but it turned out that Mrs. Taylor was much taken with Mill."[19] Like Carlyle, Roebuck assumed Mill knew nothing of women, his mother and sisters aside, before he met Harriet Taylor. Roebuck's retrospective account, recorded many decades after 1830, intimated that Harriet Taylor's response to Mill mattered far more than his response to her. We are led to suppose that if she was "taken" with him, and then acted on her feelings, he would succumb. Both Carlyle and Roebuck took it for granted that Taylor had far more say than Mill in what followed. Roebuck did not specify how much time elapsed before he "learned that an intimate acquaintance had arisen between Mill and Mrs. Taylor." That one did arise in fairly short order seems clear. The initial phase of this acquaintance must have been launched by Harriet Taylor. One need not go all the way with Carlyle and Roebuck on the subject of Mill's innocence to conclude that he could not have taken the lead. Control over the structure of the relationship, from start to finish, rested principally with Harriet Taylor, whose force of will Mill could not begin to answer.

The earliest surviving documentation attesting to a romantic friendship between John Mill and Harriet Taylor inferentially derives from August 1832. A letter from Mill, written in French and bearing no date, refers to flowers Mill had gathered in the New Forest with Harriet Taylor in mind. Mill's summer 1832 walking tour in Hampshire, Sussex, and the Isle of Wight culminated in the New Forest toward the end of the first week of August. Either upon his return, or shortly thereafter, he learned of Harriet Taylor's decision to end their relationship. The communication that passed between them on this occasion was carried out through an intermediary, very probably Eliza Flower. Mill's gallant reply delicately conveyed the heartache caused by this decision, while honorably acknowledging the duty it imposed upon him. At the same time, he would not fully embrace the idea that a conclusive break was taking place.

> Benie soit la main qui a tracé ces caractères! Elle m'a écrit—il suffit; bien que je ne me dissimule pas que c'est pour me dire un éternel adieu.
> Cette adieu, qu'elle ne croie pas que je l'accepte jamais. Sa route et la mienne sont séparées, elle l'a dit: mais elles peuvent, elles doivent, se rencontrer. A quelqu'époque, dans quelqu'endroit, que ce puisse être, elle me trouvera toujours ce que j'ai été, ce que je suis encore.
> Elle sera obéie: mes lettres n'iront plus troubler sa tranquillité, ou verser une goutte de plus dans la coupe de ses chagrins. Elle sera obéie, par les motifs qu'elle donne,—elle le serait quand même elle se serait bornée à me communiquer ses volontés. Lui obéir est pour moi une nécessité.
> Elle ne refusera pas, j'espère, l'offrande de ces petites fleurs, que j'ai apportées pour elle du fond de la Nouvelle-Forêt. Donnez-les lui s'il le faut, de votre part.[20]

[Blessed be the hand which traced these characters! She writes to me—that is enough: although I do not pretend that it is not to say goodbye to me forever.
She must not believe that I accept such a farewell. Her way and mine have separated, as she says. But they can, they ought to, meet again. At whatever time, in whatever place that may be so, she will find me always the same as I have been, as I am still.
She shall be obeyed; no further letters of mine shall disturb her peace, or pour one extra drop into her cup of sorrows. She shall be obeyed, for the reasons which she gives,—she would have been, even if she had confined herself to telling me her wishes. To obey her is for me a necessity.
She will not refuse, I trust, the offering of these little flowers, which I have brought for her from the depths of the New Forest. Give them to her, if necessary, as if they came from you.]

Harriet Taylor did not want to give up Mill, and she was soon able to persuade herself that such a sacrifice could be avoided. Her attempted withdrawal had presumably arisen from John Taylor's learning of her feelings for Mill. There is nothing to suggest that Harriet Taylor ever felt guilty about the feelings themselves. Her love for Mill, as she understood it, was pure and ennobling. Although she wondered whether the intensity of Mill's emotional investment in the relationship matched her own, she knew that his love for her was untainted by base motives. She would not be unfaithful to her husband, and Mill would not want her to be. Still, she could see that she had hurt John Taylor, from whom it could not be concealed that his wife loved another better than she loved him. He was not only her husband but the father of her three young children. He had done all he could to make her happy. She had inadvertently caused him pain, and she had tried to repair the damage by breaking with Mill. The break did not last, for she had found that the pain of the severed lovers, perhaps one of them in particular, was more than they should have to bear, given the purity of their feelings, intentions, and actions.

In October 1832 Mill spent a week walking in Cornwall. The Taylors, during the same month, toured North Wales by carriage. When the latter returned to London they moved from their Finsbury residence to a grander dwelling, 17 Kent Terrace, Regent's Park. Mill was a frequent visitor, as can be seen from a letter he wrote Fox in May 1833. "I knew not that you were to be in K.T. on Wednesday, and I seldom go there without some special reasons on that day of the week, for as it cannot be right in present circumstances to be there *every* evening, none costs so little to give up as that in which there is a much shorter time and only in the presence of others."[21] What Mill thought the right number of evenings to be "in present circumstances" is not revealed; Fox, it seems, thought his chances of finding Mill at Kent Terrace on a Wednesday evening as good as on any other evening of the week. John Taylor was a highly obliging sort of husband. The craving John Mill and Harriet Taylor had for each other's company was such that in June of the same year Mill could not stay away, though the evening in question happened to be a Wednesday. He reported to Fox: "As for *me*, I am going to K.T. today, despite its being Wednesday."[22]

The meetings at Kent Terrace were not of a clandestine character. Did they sometimes seek other venues, ones that would permit a greater degree of intimacy than Kent Terrace could accommodate? Harriet Taylor wrote to John Mill, probably around this time (the letter is undated): "Yes dear I will meet you, in the chaise, some where between this and Southend—the hour will depend on what your note says to-morrow (that is supposing the chaise is to be had of which there is very little doubt.) . . . Bless thee—to-morrow will be delightful & I am looking to it as the very greatest treat."[23] Not all was above board, and the illicit element primed the adrenal glands.

Such assignations also signaled the gap between the intimate world they sought to create for themselves and the conventions governing respectable society. The latter forbade sighs of love passing between a married woman and a man not her husband. John Mill and Harriet Taylor ruled out the possibility that any blame could be justly affixed to their conduct. The fault lay with the institution of marriage and the subjugation of women it enjoined. Fairly early in their relationship, probably soon after they had resumed seeing one another in late 1832 or early 1833, Harriet Taylor asked Mill to write down his thoughts on marriage and divorce.[24] As the request came from "She to whom my life is devoted," he naturally complied.[25]

Mill held advanced views on the subject of equality between the sexes before he met Harriet Taylor. In the *Autobiography* he notes that these opinions "were among the earliest results of the application of my mind to political subjects, and the strength with which I held them was, as I believe, more than anything else, the originating cause of the interest she [Harriet Taylor] felt in me."[26] Mill would have been reading Bentham when he first applied his "mind to political subjects." Bentham maintained that the interests of women ought to count for no less than those of men in any just understanding of "the greatest happiness principle." A passage from his *Introduction to the Principles of Morals and Legislation* (1780) would not have been out of place in J.S. Mill's *Subjection of Women*:

> In certain nations, women, whether married or not, have been placed in a state of perpetual wardship: this has been evidently founded on the notion of decided inferiority in point of intellects on the part of the female sex, analogous to that which is the result of infancy or insanity on the part of the male. This is not the only instance in which tyranny has taken advantage of its own wrong, alleging as a reason for the domination it exercises, an imbecility, which, as far as it has been real, has been produced by the abuse of that very power which it is brought to justify.[27]

James Mill, to be sure, did not join Bentham in deprecating this injustice. Yet, alongside the elder Mill's disdainful treatment of Mrs. Mill and his readiness to deny women the suffrage in the *Essay on Government* should be set his apparent expectation that the instruction John Mill gave his siblings should make no significant distinction between brothers and sisters.[28] Moreover, James Mill let stand his son's criticism of the *Edinburgh Review*'s canvassing of the "female character" when father and son went after the leading Whig quarterly in 1824. The brand of morality inculcated by the *Edinburgh*, the younger Mill noted, praised men for being energetic, independent, and self-sufficient. "In a woman, helplessness, both of mind and of body, is the most admired of attributes." A man without courage was thought contemptible; a woman without courage was thought "amiable."

> To be entirely dependant upon her husband for every pleasure, and for exemption from every pain; to feel secure, only when under his protection; to be incapable of forming any opinion, or of taking any resolution without his advice and aid; this is amiable, this is delicate, this is feminine: while all who infringe on any of the prerogatives which man thinks proper to reserve for himself; all who can or will be of any use, either to themselves or to the world, otherwise than as the slaves and drudges of their husbands, are called masculine, and other names intended to convey disapprobation. Even they who profess admiration for instructed women, not unfrequently select their own wives from among the ignorant and helpless.[29]

James Mill knew a number of well-instructed women—Sarah Austin, Harriet Grote, and Lady Bentham among them—for whom he may well have professed admiration. He had chosen as his wife a woman incapable of evoking his respect. Did his eldest son wittingly deliver an indirect rebuke to his father when offering this last observation on the foolishness of some enlightened men? His father had been foolish, but he was also enlightened enough to appreciate the good sense in what his son had to say on this subject.

The youthful John Mill independently concluded that the law should not deny women the rights and opportunities it gave to men. He says in the *Autobiography* that he strongly dissented from James Mill's contention "that women may consistently with good government, be excluded from the suffrage, because their interest is the same with men." He apparently expressed this dissent to his father, who replied that he had not "intended to affirm that women *should* be excluded, any more than men under the age of forty." These exclusions had formed part of a discussion that aimed to establish "the utmost limit of restriction" consistent with maintaining "the securities for good government." James Mill had not argued that such restrictions ought to be applied. John Mill, along with his associates of the early 1820s (over whom he exercised considerable intellectual sway), granted that his father's argument should be so understood. At the same time, they maintained (according to John Mill) that "the interest of women" was no more "included in that of men" than "the interest of subjects" was "included in that of kings." They held "that every reason which exists for giving the suffrage to anybody, demands that it should not be withheld from women." Bentham too entered the discussion, "wholly on our side."[30]

We cannot know whether John Mill's "strong convictions on the complete equality in all legal, political, social and domestic relations, which ought to exist between men and women" were "the originating cause of the interest she [Harriet Taylor] felt" in him.[31] They surely played a part. Such a straightforward characterization of these convictions might lead one to think that the issue of marriage and divorce would be swiftly and decisively dealt with by Mill in the essay he drafted at the urging of Harriet Taylor. Not so.

Neither marriage nor divorce fared well in the essay. The "law of marriage as it now exists," he declared, "has been made *by* sensualists, and *for* sensualists, and *to bind* sensualists."[32] Most men being voluptuaries, the indissolubility of marriage had long been advantageous to women. (Before 1857 divorce could only be secured through an Act of Parliament, a prohibitively expensive recourse for all but the most wealthy. Judicially authorized separations, which did not allow for remarriage, were the purview of the ecclesiastical courts.) The general run of men married to satisfy their taste for a particular woman. Once this appetite had been quenched, their natural inclination was to abandon the woman in question. The "irrevocable vow" they had taken countered this inclination by creating a permanent bond between man and woman. The mere fact of living together as husband and wife encouraged the husband to develop "some feeling of regard and attachment" for his spouse. Moreover, the wife gained what often mattered to her most—security against her children being taken from her. Most women sought in marriage a stable home and a respectable status. When these had been acquired, they wished to keep what they had. Mankind at large, at least as currently constituted, could not grasp what individuals of a higher nature readily understood: "the absurdity and immorality of a state of society and opinion in which a woman is at all dependent for her social position upon the fact of her being or not being married."[33] This dependence stemmed from the formative influences of education and custom. From the earliest age girls learned that they were meant to be wives and mothers, and that the functions they performed in these roles determined their usefulness to society. Responsibility for their material support was to rest with the male head of the household. "The indissolubility of marriage," Mill observed, was "the keystone of woman's present lot, and the whole comes down and must be reconstructed if that is removed."[34]

The fundamental question, he then asserted, was "not what marriage ought to be, but . . . what woman ought to be." Marriage between equals would be something very different from what marriage had been hitherto. Mill insisted that the progress of civilization had rendered ever less important the only natural inequality distinguishing the sexes, "bodily strength." The obstacle to converting natural equality into social equality lay "in artificial feelings and prejudices." A woman should not be dependent on a man by virtue of her being a woman, any more than a man should be dependent on a woman by virtue of his being a man. Legitimate dependence of one party on the other could arise, but only when spawned by the affections and when renewed perpetually "by free and spontaneous choice." As matters stood, however, a woman's position, "in nine cases out of ten, makes her either the plaything or slave of the man who feeds her; and in the tenth case, only his humble friend."[35] The emancipation of women from this position hinged on their receiving an education fit for the achievement of economic independence.

Did Mill expect that most women would enter the labor market, once given the means to earn their livelihood? He did not think they would, or that they should. Doubling the number of people competing for employment would place tremendous burdens on an already crowded labor market. Here Mill seemed to be implying that few women were currently in paid employment—he knew, of course, that the laboring classes included a sizeable number of women who had to work for wages in order to survive. If matters were arranged to suit his wishes, the female portion of the labor force would in fact shrink. "In a healthy state of things, the husband would be able by his single exertions to earn all that is necessary for both; and there would be no need that the wife should take part in the mere providing of what is required to *support* life: it will be for the happiness of both that her occupation should rather be to adorn and beautify it."[36]

Mill had no clue that his vision of womanhood had been infiltrated by "artificial feelings and prejudices." His lofty and idealistic conception of woman's chief vocation was firmly anchored in the culture of his age. "The great occupation of woman should be to *beautify* life: to cultivate, for her own sake and that of those who surround her, all her faculties of mind, soul, and body; all her powers of enjoyment, and powers of giving enjoyment; and to diffuse beauty, and elegance, and grace, everywhere." If her nature were so active as to require employment beyond this "great occupation," the world would not fail to provide it. Mill perhaps had a particular couple in mind when he noted how this unexhausted energy would "naturally" be put to use:

> If she loves [and how could it be otherwise for one who everywhere dispersed "beauty," "elegance," and "grace"], her natural impulse will be to associate her existence with him she loves, and to share *his* occupations; in which if he loves her (with that affection of *equality* which alone deserves to be called love) she will naturally take as strong an interest, and be as thoroughly conversant, as the most perfect confidence on his side can make her.[37]

A woman with a life such as this had fulfilled "the end of her existence." Yet, even in a world where legal and social equality prevailed between the sexes, many women would be denied this ultimate fulfillment. The parents of girls had to ensure that their daughters had "the power of gaining their own livelihood."[38] Although there should be no distinction in the employments for which daughters and sons were to be fitted, nature could be relied upon to make its own distinction. Those occupations "which partake most of the beautiful, or which require delicacy and taste rather than muscular exertion, will naturally fall to the share of women: all branches of the fine arts in particular."[39]

Having said this (and more), Mill was nearly ready to pronounce judgment on what the law of marriage should look like. A just law of marriage presupposed that women were free to stay unmarried, a freedom founded on their ability to support themselves. In such circumstances, those who chose to marry would not need to remain married for reasons of subsistence. In many instances, the dissolution of a marriage would increase the sum of happiness of the parties concerned, if only by lessening the sum of their unhappiness. A large number of marriages inevitably scored poorly on happiness: "the failures in marriage are such as are naturally incident to a first trial: the parties are inexperienced, and cannot judge. Nor does this evil seem to be remediable."[40] Allowing parents to decide who would be a fit spouse for their son or daughter could not answer this problem. The motives fathers and mothers had in promoting or discouraging a particular match could be far removed from the true interests of those who had the greatest stake in the outcome. Given the chances of a marriage turning out badly, strong grounds existed for permitting divorce.

Those in favor of making divorce a readily available option nonetheless had to consider the negative effects of adopting this policy. If people could marry, divorce, and remarry with ease, they would have much less incentive to choose well in the first place. And Mill granted cogency to the position "that the first choice should be, even if not compulsorily, yet very generally, persevered in."[41] Serial "trials for happiness, and repeated failures," he maintained, "have the most mischievous effect on all minds."[42] The blighting of hope engendered by such recurring disappointments would sow despair among those who aspired to a virtuous life. The effect of multiple matrimonial trials and errors on the more common stock of humanity would be a sinking into moral depravity. Mill, moreover, was persuaded that the interests of children tended to be far better served if the family stayed intact. Children, he stated, "are wholly dependent for their happiness and for excellence upon their parents." In the vast majority of cases, that happiness and excellence "*must* be better cared for ... if their parents remain together."[43] Mill also introduced a further consideration touching upon the issue of divorce: the gap between the immoderate expectation of happiness held by most people and their limited capacity for experiencing it. When marriage did not yield the degree of happiness they had unreasonably anticipated, they identified their discontent with the shortcomings of their spouse. Were divorce easy to procure, they might be tempted to seek a different partner, whom they mistakenly imagined would give greater satisfaction. Better that the original marriage be sustained—union over time had a way of diminishing disappointment, and the parties concerned would commonly enjoy as much happiness as they were capable of knowing. The corrosive effects of repeated marital experiments conducted upon faulty premises would do serious harm.

Having discharged his obligation to throw on the scale all that could be said against giving ready access to divorce, Mill declared that these arguments rested on an inapt assumption: that "the choice lies between the absolute interdiction of divorce, and a state of things in which the parties would separate on the most passing feeling of dissatisfaction."[44] Legal obstacles need not be placed in the way of divorce in any society opposed to "promiscuous intercourse." Be divorce ever so freely available, the force of opinion would powerfully discourage a casual attitude toward the ending of one marriage and the starting of another. Those adopting such an attitude, women particularly, would feel society's scorn. Mill proclaimed his conviction that "the first choice would almost always, especially where it had produced children, be adhered to, unless in case of such uncongeniality of disposition as rendered it positively uncomfortable to one or both of the parties to live together, or in case of a strong passion conceived by one of them for a third person."[45] Mill did not mean to say that a dismal lack of congeniality or the rise of such a passion should necessarily bring about the dissolution of a marriage. His position, instead, was that the union should not "be forcibly preserved" where either of these circumstances obtained. Marriage, like all other contractual relations, should now "depend for its continuance upon the wishes of the contracting parties." Although the advance of civilization had made women "ripe for equality," this equality could not be realized without reasonable legal provision being made for the dissolubility of marriage. By "no other means can the condition and character of women become what it ought to be."[46]

Harriet Taylor had asked Mill for "a written exposition" of his views on the subject of marriage and divorce, and a written exposition she got. As a private communication his miniature treatise nonetheless has an eccentric cast. Why tread such a tortuous path to reach a judgment that marriage, as a contractual agreement, should bind the parties only so long as they wished to be bound? Mill surely grasped that the person soliciting his opinion already knew her own mind on this matter. She had not asked for his thoughts in order to consider the wisdom of making them her own. She wanted to discover whether he shared her understanding *and* depth of feeling regarding the injustice inflicted on women by society and the law. Mill's analytically overwrought response displayed the effects of his intellectual training and his affinity for the habits of mind it instilled. In this instance, he took advantage of his capacity to function in a way that Harriet Taylor could not. He could feel perfectly justified in handling the issues as he did, and his recipient could not reasonably expect him to adopt a different mode in treating them. Injecting into the discussion emotional considerations bearing explicitly on the specific circumstances of John Mill and Harriet Taylor would compromise the integrity of the analysis. The latter person, in any event, was exempt from the precepts that should regulate the conduct of ordinary people, as Mill made clear in an elaborately deferential introductory section of his essay.

> If all resembled you, my lovely friend, it would be idle to prescribe rules for them. By following their own impulses under the guidance of their own judgment, they would find more happiness, and would confer more, than by obeying any moral principles or maxims whatever; since these cannot possibly be adapted beforehand to every peculiarity of circumstance which can be taken into account

by a sound and vigorous intellect *worked* by a strong *will*, and guided by what Carlyle calls "an open loving heart."[47]

Having conceded so much to Harriet Taylor's superior nature, Mill could thereafter don the mantle of the impartial investigator and engage in a nuanced examination of the problem. The exemption, and the examination it prefaced, significantly reduced his personal liability for anything said that might offend the sensibilities of his "lovely friend."

Mill's essay on marriage and divorce fulfilled his part of the pact. Harriet Taylor executed her part in a short essay as brisk and direct as Mill's was cautious and circuitous. Brisk and direct, in this case, should not be confounded with clear and coherent. One never doubts that Mill constructed his sentences with a view to their yielding a graspable meaning to the reader. It is alarming to think that the opening sentence of Harriet Taylor's commentary must have made sense to its writer. "If I could be providence to the world for a time, for the express purpose of raising the condition of women, I should come to you to know the *means*—the *purpose* would be to remove all interference with affection, or with any thing which is, or which even might be supposed to be, demonstrative of affection."[48] A paragraph so begun did not have to worry about losing its way thereafter.

Harriet Taylor, in fact, did not expect that good results would follow from the immediate elimination of all restraints on the expression of affection so long as women remained in a state of sheer ignorance and total dependence. The consequences of such a radical change "would probably be mischievous." The thing most needful in existing circumstances was "to give women the desire to raise their social condition." But who was to do the giving? Presumably not men, all of whom, "with the exception of a few lofty minded, are sensualists more or less." There was "equality in nothing," least of all in the experience of enjoyment: "all the pleasures such as there are being mens [sic], and all the disagreables [sic] being womens [sic]." Harriet Taylor allowed for the possibility that male sensualism was not innate, but instead a product of "the habits of freedom and low indulgence in which boys grow up." If the problem of legal and social inequality could be overcome, "every pleasure [for both men and women] would be infinitely heightened both in kind and degree." This could never happen in a society where women were conditioned to believe that their sole object was "to gain their living by marrying." A living might so be won, but marriage seldom brought "real sympathy or enjoyment of companionship."

> The woman knows what her power is, and gains by it what she has been taught to consider "proper" to her state—The woman who would gain power by such means is unfit for power, still they *do* use this power for paltry advantages and I am astonished it has never occurred to them to gain some large purpose: but their minds are degenerated by habits of dependance [*sic*].[49]

Even these observations scarcely prepared the ground for Harriet Taylor's breathtaking assertion that once "the whole community is really educated, tho' the present laws of marriage were to continue they would be perfectly disregarded, because no one would marry."[50]

Society could not wait upon the arrival of this happy day before supplying a remedy for the misery caused by marriage. Harriet Taylor's prescription: the opportunity for all to obtain an inexpensive divorce without being required to justify the application. She allowed that some time should elapse—two years struck her as reasonable—between initiation of the action and the issuing of a final decree dissolving the marriage. Provided the suit was not withdrawn during this interval, complete certainty of its success must be assured. The granting of the decree should free the parties to marry again (such a foolish choice would go on being made in a benighted society).

Implicit in Harriet Taylor's discussion was the assumption that divorce would not be uncommon, and that people would come to understand that when one party sought to end a marriage the other party should have no interest in resisting this course. In the present social order, young women entered "what is called a contract perfectly ignorant of the conditions of it, and that they should be so is considered absolutely essential to their fitness for it!"[51] The rectifying of this ignorance after the wedding ceremony led some wives to wish they had remained single. If divorce were a readily available option, a proportion of these women would exercise it. "Whoever would take the benefit of a law of divorce must be those whose inclination is to separate and who on earth would wish another to remain with them against their inclination?"[52]

An odd question, perhaps, coming from one who had just depicted nearly "all men" as "sensualists more or less." Harriet Taylor's proposals on divorce were meant for a society in which most people failed to see that marriage, in principle, was a bad idea. In such a society most men would continue to be sensualists, and "all the pleasures" would continue to be theirs, "all the disagreables [*sic*] and pains being womens [*sic*]." Evidently, there were quite a few males on this earth who had no difficulty accommodating this disparity, and who would wish to see the arrangement preserved whatever the desire of the other party might be.

Taking civilization as it currently stood, what would Harriet Taylor recommend if she had full power to mandate changes in the law? A law of divorce, in a legal system designed by Taylor, would be superfluous. That system would give to women all the "rights and privileges, civil and political" that belonged to men, and no laws relating to marriage would be called for. Women would have responsibility for supporting their children. Understanding this to be the case, they would refrain from having children unless they knew they could provide for them. No disabilities would be artificially imposed on women with respect to economic opportunity. "Women would have no more reason to barter person for bread, or for any thing else, than men have—public offices being open to them alike, all occupations would be divided between the sexes in their natural arrangement. Fathers would provide for their daughters in the same manner as for their sons."[53] What it meant for "fathers" to make such provision in a world where women had responsibility for meeting the subsistence needs of their children was not revealed. Would fathers end up providing for their children because the children in question would have no mothers?

The penultimate and final paragraphs of Harriet Taylor's composition invite speculation about the thought processes at work. The former tersely elaborated on the subject of children, mingling several propositions (within the confines of a single run-on sentence): first, the difficulties raised by the issue of divorce concerned its implications for the well-being of the children affected; second, on the plan proposed by Taylor the interest of women would lie in their not becoming mothers; third, in existing circumstances, a woman sought security through having "children as so many *ties* to the man who feeds her."[54] Taylor developed none of these points before shifting abruptly into a lyrical register for her last paragraph, in which she delineated "the true and finest meaning" of "*Sex*." In this meaning the five senses acted as the servants of an end that wholly transcended the physical. "Are we not born with the *five* senses, merely as a foundation for others which we may make by them—and who extends and refines those material senses to the highest—into infinity—best fulfils the end of creation." Mill himself, "the apostle of all loftiest virtue," must teach mankind "that the higher the *kind* of enjoyment, the *greater* the *degree*."[55]

The kind of enjoyment Harriet Taylor aspired to, an enjoyment that refined the "material senses . . . into infinity," did not carry a risk that offspring might issue from the experience. Marriage, sexual intercourse, children—the attitude Harriet Taylor evinced toward each and all could be taken as an exemplary demonstration of the power of association. She did not need to manifest a capacity for lucid and consecutive exposition to transmit forcefully the intensity of her aversions. Harriet Taylor's aspirations and aversions would have serious practical effects.

In September 1833 Harriet Taylor left London for Paris, her husband and three young children staying behind at Kent Terrace. John Taylor, made miserable by his wife's refusal to give up Mill, had proposed a trial separation. He would provide for his wife's travel expenses. Neither party wished to provoke a permanent break. As Mill later explained to W.J. Fox, John Taylor had taken the initiative. Seeing that his wife would not undertake "to give up either the feeling, or the power of communication with me—unless she did so, it was *Mr. Taylor's* wish, and seemed to be necessary to his comfort that she should live apart from him."[56] Taylor may well have calculated that his wife, finding separation from her children and familiar way of life unendurable, would grasp the full import of what she stood to lose by not renouncing her connection with Mill. He hoped for her return to Kent Terrace on terms that would restore some semblance of normality to their domestic life. Harriet Taylor's unwillingness to sever her ties with Mill had produced enormous emotional strain on both husband and wife. For all her self-command, Harriet Taylor struggled to hold at bay the acute anxiety created by the predicament she had brought on herself and those nearest to her. No relief could be had so long as she remained at Kent Terrace. She could tolerate a short-term experiment that removed her from this setting, with her husband's sanction and support, and that offered her the prospect of seeing John Mill after she got settled in Paris.

Uncertainty and unhappiness naturally vexed John Mill too in the months preceding the Taylors' trial separation. His letters to Carlyle during the first half of 1833, without hinting at the cause of his malaise, made plain his somber cast of mind. In March he stated: "You see it is cold comfort which I can give to any who need the greatest of comforts, sympathy in moments of dejection; I, who am so far from being in better mental health than yourself, that I need sympathy quite as much, with the added misfortune that if I had it, it could do me no good."[57] A month later Mill told Carlyle that he had let himself become "paralysed more than I should, during the last month or two by these gloomy feelings, though I have had intervals of comparative brightness but they were short."[58]

The paralysis of which Mill spoke had afflicted John and Harriet Taylor as well; the plan for Harriet to leave Kent Terrace was designed to produce a remission, one John Taylor hoped would lead to a remedy consistent with his own interests and those of his children. John Mill had hopes of his own, although he apparently had difficulty determining precisely what they were. As of 5 September he learned of Harriet Taylor's impending departure for Paris. On that date Mill wrote to Carlyle, explaining that he could not carry into effect his projected autumn visit to Craigenputtock, the small farm in southwestern Scotland where the Carlyles had taken up residence.

> There were about twenty chances to one that I should [have gone to Craigenputtock], but it is the twenty-first which has taken effect in Reality. I was mistaken, too, when I said that if I went not to Craigenputtoch I should go nowhere: I am going to Paris: the same cause which I then thought, if it operated at all, would operate to keep me here, now sends me there. It is a journey entirely of duty; nothing else, you will do me the justice to believe, would have kept me from Craigenputtock after what I have said & written so often: it is duty, and duty connected with a person to whom of all persons alive I am under the greatest obligations.[59]

For some time Mill had understood that the wretched state of affairs at Kent Terrace might produce a dramatic development that would involve Harriet Taylor leaving her husband. The likelihood of this occurring by late summer or early autumn had struck him as fairly small. Events had swept Mill along, neither the timing of the separation nor Harriet Taylor's destination falling within his purview. He did not resent this lack of control over what was happening; he did worry about defects in his own character that kept him from fully knowing and acting upon his own will.

Mill's emotional reserve had troubled Harriet Taylor for some time. The climactic events of early September 1833, which seemed to call for a decisive response on his part, moved him to tell her of his deep self-doubt. The letters Mill wrote Harriet Taylor during this period have not survived. A long letter she wrote, postmarked 6 September, was preserved, and it alluded to what he had revealed.

> I am glad that you have said it—I am *happy* that you have—no one with any fineness or beauty of character but must feel compelled to say *all*, to the being they really *love*, or rather with any *permanent* reservation it is *not* love,—while there is reservation, however little of it, the love is just *so much* imperfect. There has never, *yet*, been

entire confidence *around* us. The difference between you and me in that respect is, that *I* have always *yearned* to have *your* confidence with an intensity of wish which has *often*, for a time, swallowed up the naturally stronger feeling. The affection itself—you have not given it, not that you *wished* to reserve—but that you did not *need* to give—but not having that need of course you had no perception that I had & so you have discouraged confidence from me till the habit of *checking first thoughts* has become so strong that when in your presence timidity has become almost a *disease* of the nerves. It would be absurd only it is so painful to notice in myself that every word I ever speak to you is detained a second before it is said till I'm quite sure I am not by implication asking for your confidence. It is but that the only being who has ever called forth all my faculties of affection is the only in whose presence I ever felt constraint. At times when that has been strongly felt *I* too have doubted whether there was not possibility of dissappointment [*sic*]—that doubt will never return. You can scarcely conceive dearest what *satisfaction* this note of yours is to me for I have been depressed by the fear that what I wd most wish altered in you, *you* thought quite well of, perhaps thought the best of your character. I am quite sure that want of energy *is* a defect, would be a defect if it belonged to the character, but that thank Heaven I am sure it does not. It is such an opposite to the *sort* of character.

Yes—these circumstances *do* require greater strength than any other—the greatest—that which you have, & which if you had not I should never have loved you, I should not love you now. In this, as in all the most important matters there is no medium between the *greatest*, *all*, and none—anything less than all being insufficient there might be just as well none.

If I did not know them to be false, how heartily I should scorn such expressions, "I have ceased to will"! then to *wish*? for does not *wish* with the power to *fulfill* constitute *will*? It is false that your "strength is not equal to the circumstance in wh you have placed" yourself.—It is quite another thing to be guided by a judgment on which you can rely & which is better placed for judgment than yourself.

Would you let yourself "drift with the tide whether it flow or ebb" if in one case every wave took you further from me? Would you not put what strength you have into resisting it? Would you not *wish* to resist it, would you not *will* to resist? Tell me—for if you would not, how happens it that you *will* to love me or any *most dear*!

However—since you tell me the evil & I believe the evil, I may surely beleive [sic] the good—and if all the good you have written in the last two or three notes be *firm truth*, there is *good enough*, even for me. The most horrible feeling I ever know is when for moments the fear comes over me that *nothing* which you say of yourself [can be] absolutely relied on. That you are not *sure* even of your strongest feelings. Tell me again that this is *not*.[60]

Mill had learned to think for himself. His cultivation of the sympathies notwithstanding, he doubted he had learned to feel for himself. This wariness engendered an emotional passivity that contrasted sharply with Harriet Taylor's emotional aggressiveness. What would she have made of his telling Carlyle that his sense of duty and obligation mandated his choosing Paris over Craigenputtock? Could he not say what he wanted? Could he not know what he wanted? The question he had asked himself in the midst of his depression of 1826 had begun: "Suppose all your objects in life were realized"; when his "heart sank within him," it did so because of the gaping hole between such "objects" and genuine "wants." A suppression of the will had been essential to the survival strategy he had employed during his formative years, when self-assertion had been the enemy of self-protection. How much strength of will did it take for John Mill to suppress his will? Perhaps not a lot. By the time he met Harriet Taylor he did not lack for self-esteem. The identity of the self he esteemed, however, was largely defined by capacities and convictions—intellectual, moral, and political. Such an identity had virtually no standing when it came to issues whose bearing on intimate relationships was most trenchant. Harriet Taylor's adamantine self-esteem acted as an infinitely renewable and immensely formidable resource because of its inseparability from her emotional core (or from her idea of what constituted that core). She told John Mill: "The desire to give & to receive feeling is almost the whole of my character."[61] What an impossible task it would have been for John Mill to sum up the essence of his "character" in eight words, or eighty. Harriet Taylor would not go so far as to tell John Mill what his feelings were; she had no hesitation telling him what they meant.

She would see to it that Mill had strength equal to the circumstances in which he had placed himself. "To obey her is for me a necessity"—something more than a chivalrous conceit in Mill's case? Harriet Taylor assured him that the judgment upon which he could rely was near at hand, and that it was "better placed" than his own for knowing what the circumstances required of him. In addition to the passages already quoted, her letter included a declaration that "the *only* evil there can be for me is that you should *not* think my best your best—or should not agree in *my* opinion of my best." Of herself, Harriet Taylor had no wariness.

Harriet Taylor's letter must have bucked up Mill a good deal. He concluded that the Taylors' separation markedly improved his prospects of winning this great prize for himself. Looking back upon the month leading up to his joining Harriet Taylor in Paris, Mill cited his "hopefulness and happiness." He "felt an immense increase of the chances in [his] favour."[62]

Their Paris rendezvous began in mid-October.[63] Mill arrived in Paris believing that Harriet Taylor had cast off doubts regarding his fitness to be her partner. He discovered, to his dismay, that she still lacked confidence in his readiness to think her best his best and to agree in her opinion of what she thought best. "When I came here [Paris]," he later wrote Fox, "I *expected* to find her no more decided than she had always been about what would be best for all, but *not* to find her as for the first time I did, doubtful of what would be best for our *own* happiness." This latter indecisiveness aroused "painful feelings" in Mill.[64] He was soon able to show her that she had no cause to doubt his being agreeable, no reason to worry whether he would deem best what she deemed best. The Paris episode created a far greater measure of emotional trust. In early November Mill gave Fox and Eliza Flower, the two people who alone could be counted upon for moral support, a waft of the intimacy that had arisen during this extended period of intense physical and emotional togetherness.

> I could have filled a long letter to you with the occurrences and feelings and thoughts of any one day since I have been here—this fortnight seems an age in mere duration, and *is* an age in what it has done for us two. It has brought years of experience to us—good and happy experience most of it. We never could have been so near, so perfectly intimate, in any former circumstances—we never could have been together as we have been in innumerable smaller relations and concerns—we never should have spoken of all things, and in all frames of mind, with so much freedom and unreserve. I am astonished when I think how much has been restrained, how much untold, unshewn, uncommunicated till now—how much which by the new fact of its being spoken, has disappeared—so many real unlikenesses, so many more false impressions of unlikeness, most of which have only been revealed to me since they have

ceased to exist or these which still exist have ceased to be felt painfully. Not a day passed without removing some real & serious obstacle to happiness. I never thought so humbly of myself compared with her, never thought or felt myself so little worthy of her, never more keenly regretted that I am not, in some things, very different, for her sake—yet it is much to know as I do now; that almost all which has ever caused her any misgivings with regard to our fitness for each other was mistaken in point of fact—that the mistakes no longer exist—& that she is now (as she is) quite convinced that we are perfectly suited to pass our lives together—better suited indeed for that perfect than for this imperfect companionship. There will never again I believe be any obstacle to our being together entirely, from the slightest doubt that the experiment would succeed with respect to ourselves—not, as she used to say, for a short time, but for our natural lives. And yet—all the other obstacles or rather the one obstacle being as great as ever—our futurity is still perfectly uncertain. She has decided nothing except what has always been decided—not to renounce the liberty of sight—and it does not seem likely that anything will be decided until the end of the six months, if even *then* finally. For me, I am certain that whatever she decides will be wisest and rightest, even if she decide what was so repugnant to me at first—to remain here alone—it is repugnant to me still—but I can now see that perhaps it will be best—the future will decide that.[65]

The one obstacle, no small one, was John Taylor and the deep affection his wife continued to feel for him. Harriet Taylor's passion for Mill had only reinforced her affection for her husband, "by so many new proofs of *his* affection for *her*, & by the unexpected & (his nature considered) really admirable generosity & nobleness which he has shewn under so severe a trial." (The condescending attitude is unmistakable, Mill implying that such generosity and nobility would be less worthy of praise if evinced by a person of superior sensibilities.) Harriet Taylor's affection for her husband did not rise to the level of a need for having him around. Provided she knew "him to be happy though away from her," she "would be satisfied." Her feelings for Mill were of a different order, and she could not reconcile herself to being forever away from him. The dilemma she faced arose from John Taylor's insistence that a permanent separation from his wife would smash his happiness. She could not bear giving up her passion for Mill; neither could she live with the knowledge that she had made her husband "durably wretched."[66]

Both men had been stymied. Harriet Taylor, Mill told Fox, "believes—& she knows him better than any of us can—that it [an everlasting separation] would be the breaking-up of his whole future life—*that* she is determined never to be the cause of, & I am as determined never to urge her to it, as convinced that if I did I should fail." It should not be supposed that Harriet Taylor saw no way out of her predicament. As Mill said in his letter to Fox, she aimed to make her husband "understand the exact state of her feelings, and . . . give *him* the choice of every possible arrangement except entire giving-up [of Mill]."[67]

The trial separation did not last six months. Before the end of November 1833 the Taylors had agreed upon an arrangement that would bring Harriet Taylor back to Kent Terrace. The details, whatever Mill might have thought about the range of choice open to John Taylor, were almost certainly authored by Harriet Taylor. Once back in London, Mill kept Fox apprised of developments. From a letter he wrote Fox on 22 November it can be inferred that Harriet Taylor had found a way to get most of what she wanted (and to escape what she did not want). Mill would remain in the picture; she would return to her husband on the condition that he forsake his sexual claims and be "with her as a *friend* and *companion*"; unstated by Mill, but presumably an integral element in John Taylor's accepting this severe curtailment of his husbandly privileges, there would be no sexual intimacy between Harriet Taylor and John Mill. She had accomplished all this while at the same time convincing her husband that he had been more in the wrong than she. Mill incorporated, in his letter to Fox, a passage from a letter he had recently received from Harriet Taylor.

> I had yesterday one of those letters from Mr. Taylor which make us admire & love him. He says that this plan & my letters have given him delight—that he has been selfish—but in future will think more for others & less for himself—but he still talks of this plan being good *for all*, by which he means *me* [inferentially Harriet Taylor's emphases], as he says he is sure it will "prevent after misery" & again he wishes for complete confidence.[68]

John Taylor, the father of three young children, might have dissented from his wife's construal of what he meant by "*for all*."

Harriet Taylor and Mill strove to sustain the closeness they had achieved in Paris in the months following her return to Kent Terrace. Bringing this off while she lived once more with husband and children, and he with parents and siblings, presented difficulties they had been spared during their time together in Paris. In the midst of the emotional turmoil of those Paris days, Mill wrote that he thought Harriet Taylor "on the whole far happier than I have ever known her."[69] The Harriet Taylor with whom he dealt in 1834 was, on the whole, less happy. A letter she probably wrote in the early part of this year conveyed sadness, with no expectation of relief. "*Happiness* has become for me a word without meaning—or rather the meaning of the word has no existence in my beleif [sic] . . . [T]he *most* this world can do for me is to give present enjoyment sufficient to make me forget that there is nothing else worth seeking."[70] They saw each other when they could, usually several times each week. Their devotion did not falter, but their spirits sometimes sagged. "I don't know why I was so low when you went this morning," she lovingly confided. "I was *so* low–I could not bear your going my darling one; yet I should be well enough accustomed to it by now."[71] A pitiable state she was in; not so pitiable, however, as to keep her from putting on a brave face for the benefit of her beloved (she relied upon Mill to discern the effort this required). "I cannot express the sort of dégout I feel whenever there comes one of these sudden cessation [sic] of life—my spiritual life—being much with you—but never mind—it is all well & right & and very happy as it is. only [sic] I long unspeakably for Saturday."[72]

W.J. Fox and Eliza Flower would have known why Harriet Taylor longed "unspeakably for Saturday." Mill assured Fox, in a letter of February 1834, that he (Mill) knew "all about the Saturday scheme, & in any way if it takes effect I hope to have a share in it."[73] The "scheme" was designed to give Mill and Harriet Taylor an opportunity to be a "couple" while in the company of sympathetic friends. Mill had earlier lamented to Fox: "I do not see half enough of you—and I do not, half enough, see *anybody* along with her—that I think is chiefly what is wanting now—that, and other things like it."[74] A mutual sense of social isolation gave added value to the Mill/Taylor-Fox/Flower coalition.

Among those who knew Mill fairly well, Thomas Carlyle was probably one of the last to learn of his involvement with Harriet Taylor. Carlyle left Craigenputtock in May 1834 to search for a house in London for himself and his wife. News of the relationship reached him within days of his arrival. A long letter to Jane Carlyle, dated 21 May, informed his wife of what he had heard about Mill's personal affairs: "Mrs Austin had a tragical story of his having fallen *desperately in love* with some ill-*married* philosophic Beauty (yet with the *innocence* of two sucking doves), and being lost to all his friends and to himself, and what not." The subject also came up in a conversation he had with Charles

Buller, whose parents had employed Carlyle to serve as their son's tutor during the first half of the 1820s. Buller had subsequently made a name for himself in the Cambridge Union. Upon settling in London to prepare himself for a legal and political career, he joined the London Debating Society, within which he ably reinforced the ranks of the radicals. John Mill respected Buller's politics and abilities; Buller admired Mill's mind. Clever, irreverent, and fun-loving, Buller was not one to take a solemn view of Mill's situation. Carlyle noted: "Buller also spoke of it; but in the comic vein."[75] Behavior observed, no less than gossip consumed, may have contributed to Sarah Austin's alarm and Charles Buller's antics.[76]

J.A. Roebuck's reminiscences state that Mill brought Harriet Taylor to an evening party hosted by Charles Buller's mother. "I saw Mill enter the room with Mrs. Taylor hanging upon his arm. The manner of the lady, the evident devotion of the gentleman, soon attracted universal attention, and a suppressed titter went round the room."[77] Mill, it seems, had decided he could take his beautiful, witty, intelligent friend to social gatherings without provoking criticism or derision. He wanted her to be admired, and he wanted others to know of his good fortune in being associated with her. Neither Mill nor Taylor saw anything dishonorable in this association. They were not slaves of their "animal appetites"; they "disdained . . . the abject notion that the strongest and tenderest friendship cannot exist between a man and a woman without a sensual relation, or that any impulses of that lower character cannot be put aside when regard for the feelings of others, or even when only prudence and personal dignity require it."[78] Yet, even innocent sucking doves could not fail to notice, or resent, the signs of disapproval.

A letter from Roebuck to his fiancée, Henrietta Falconer, indicates that he broached the matter with Mill as early as mid-September 1833. Roebuck often visited Mill at India House. On 17 September their conversation turned to a consideration of Roebuck's prudence.

> Whereupon I could not help saying, "Well John by this time I had hoped you would have a high respect for my prudence." "Why no", was his answer, "not for your prudence, but your audacity—you do things the most audacious, and succeed like a prudent person." I laughed at his nice distinction: and quickly turned the tables on him—not so however that he could draw any inference as to his own prudence. The matter was a difficult one—and I hardly knew how to proceed. However Miss Martineau's Loom and Lugger [a volume in Martineau's multivolume *Illustrations of Political*

Economy, a series that appeared in the years 1832–34] was lying on his table, and I commenced after this fashion. "Now John do not stop me till I have got to the end of my story (and I kept walking up & down his room not looking at him, in fact looking at my feet). You must not give me any confidence—no confession will I hear—but you know this lady [Martineau] has been tattling about you." He assented. "You know on what subject", he nodded aye "Well I am going to speak on that score"—and he turned as pale as death . . . I could see enough of him to see that. Poor fellow. I went on with my sermon—gave him sage counsel, and endeavoured to wean him from an ill-judged passion. A more thoroughly pure, and exalted one never was felt by any one, but unhappily to feel it, is not wise—not permitted. He believes, I discovered rather by what he did not say, than what he did, that he may indulge it, without harm to anyone. But he is killing himself . . . I left him, sad myself to see his giant mind thus torturing and enslaving him. I hope however for the best.[79]

Even before going to Paris in October, Mill would have had cause to suspect that none of his old friends considered his passion for Mrs. Taylor well-judged. That he should confide only in Fox was understandable. Fox's predicament was in some ways more harrowing than Mill's. His peculiar domestic arrangements, together with his advocacy in the *Monthly Repository* of divorce on grounds of irreconcilable differences, had created much restiveness within his congregation. His unhappy wife wanted some of his congregants to know of the unjust treatment she had received at the hands of their pastor and his ward. Fox's affairs were reaching a critical turn. A reciprocal sympathy between Mill and Fox had evolved during the early 1830s owing to the circumstances that had made each important to the other. All the same, Fox himself took issue with the terms of the triangular compact contrived by Harriet Taylor. He had reckoned that the Taylors' trial separation would lead to a definitive break and to an open avowal of partnership between John Mill and Harriet Taylor. Fox was moving toward a complete break with his wife, and he wished for passion to triumph over convention. Harriet Taylor's return to her husband had disappointed him. Mill could speak of his problems to Fox in a way that he could speak to no other man; but he could not expect his listener to sanction the compromise that Mill himself was finding hard to live with. A letter he wrote Fox in late June 1834 attested to his frustration, and alluded to the limits of Fox's sympathy.

> *Our* affairs have been gradually getting into a more & more unsatisfactory state—and are now in a state, which, a very short time ago, would have made me quite miserable but I am altogether in a higher state than I was & better able to conquer evil & to bear it . . . I have not spoken much to you about our affairs lately, as I did while she was away [in Paris]; partly because I did not so much *need* to give confidence & ask support when she was with me, partly because I know you disapprove & cannot enter with the present relation between her & me & him. but [*sic*] a time perhaps is coming when I shall need your kindness more than ever—if so, I know I shall always have it.[80]

Around this time Fox needed more kindness and support than did Mill. By July 1834 Mrs. Fox had made enough fuss over her husband's conduct to bring demands for his resignation from some members of the South Place congregation. To his friends, Fox maintained that he had not committed adultery with Eliza Flower. Yet he regarded it beneath his dignity to deny the charge in a formal statement. On 15 August he offered to resign; in September a majority of his congregation gave him a vote of confidence and asked him to withdraw his resignation. Withdraw it he did, and a significant section of those who had opposed him thereupon left South Place Chapel.[81] They must have felt vindicated when Fox and Eliza Flower established their own household in Bayswater in January 1835.

For all his problems, Fox decisively acted upon his personal affairs. Mill could not follow suit. If his personal affairs were to be sorted out, circumstances and personalities dictated that Harriet Taylor must do the sorting. She regulated the relationship, and Mill sometimes had to scramble simply to figure out what she expected of him. A Mill letter that appears to date from the post-Paris adjustment phase of their attachment underlines his occasional perplexity. He had received from Harriet Taylor a "dear letter sweet & loving," which told of the pain caused by his shutting down her attempt to express her feelings on a matter she considered important during a walk they had recently taken. What she wrote plainly upset Mill a good deal. He noted "how very much it grieves me now when even a small thing goes wrong now that thank heaven it does not often happen so, & therefore always happens unexpectedly. As for my saying 'do not let us talk of that now' I have not the remotest recollection of my having said so, or what it was that I did not want to talk about." He assumed it must have been a matter he deemed "settled & done long ago, & therefore not worth talking any more about, a reason which you yourself so continually express for not explaining to me or telling me about impressions of yours, uncertainty about the nature of which is tormenting me" (this turning of the tables indicates how unjust Mill felt the charge of insensitivity to be). He should not be held responsible for failing to intuit what she had left unsaid. "O my own love, if you were beginning to say something which you had been thinking of for days or weeks, why did you not tell me so? why did you not make me feel that you were saying what was important to you, & what had not been said or had not been exhausted before?" Mill then located the current misunderstanding in the context of a long-running dialogue respecting Harriet Taylor's "determined resolution that there *should be* radical differences of some sort in some of our feelings." He had contested this notion, and believed he had persuaded her that such differences as existed need not make them unhappy. From her he had learned "to bear that there should be some [differences]—consisting chiefly in the want of some feelings in me which you have." Albeit bewildered by the hurt he had inadvertently caused, Mill had deftly defended himself while showing a deep concern for Harriet Taylor's emotional welfare. And it surely did him no harm to close with a tender allusion to the incapacitating effect on him of the pain she had experienced. "I feel utterly unnerved & quite unfit for thinking or writing or any business—but I shall get better, & don't let it make you uncomfortable mine own—o [*sic*] you dear one."[82] At the end of Mill's letter there appears, in Harriet Taylor's hand, the words: "my *own adored* one!"[83] He had lovingly supplied the reassurance she had coveted.

Their inconclusive situation nonetheless made it hard to rid the air of heavy emotional turbulence. A letter from Harriet Taylor to Mill, penned on paper with an 1835 watermark, categorically rejected judgments he had conveyed in a recent letter. She stated "I am not one to 'create chimeras about nothing'" and declared untrue his assertion that her character was "'the extreme of anxiety and uneasiness.'" Her circumstances, she confessed, had affected her mood and made her "morbid," anxious, and uneasy. She was afflicted in body and mind; "a better state of health," she held, would remove "those morbid & weakly feelings & views & thoughts." This letter contained numerous affirmations about what she knew and he did not: he did not know the "best" of her; he did not know the effects of the circumstances she confronted daily; he did not know that she would "make the very best" of a life with him were she to decide to leave John Taylor. The fact that she had not left her husband had nothing to do with uncertainty about whether she could be happy living with Mill. She drew back from making this choice because it would mean putting her own pleasure before her "only earthly opportunity of 'usefulness.'" To abandon her husband would be to "*spoil* four lives & injure others . . . Now I give pleasure around me, I make no one unhappy, & am happy tho' not happiest myself."[84]

This self-justification was not a response to Mill's urging her to leave John Taylor. The burden of Harriet Taylor's letter instead suggests that Mill had expressed grave doubts about their prospects for creating a successful life together should she definitively break with her husband. Moreover, he had evidently voiced concern over the impact such a break could have on his own fortunes. A man who ran off with the wife of another man could not easily repair the damage done his reputation. Mill aspired to influence his society in important ways, an aspiration that would be severely compromised should he flagrantly violate the canons of upright conduct. He risked becoming "obscure insignificant & useless" (words Harriet Taylor placed within inverted commas).

The longer Harriet Taylor reflected on these words, the more she fumed. Her first reply to Mill's letter was written on a Tuesday evening. The next day she tore into him.

> Good heaven have *you* at last arrived at fearing to be "*obscure & insignificant*"! What *can* I say to that but "by all means pursue your brilliant and important career". Am *I* one to choose to be the cause that the person I love feels himself reduced to "obscure & insignificant"! Good God what has the love of two equals to do with making obscure & insignificant if ever you *could* be obscure & insignificant you *are* so whatever happens & certainly a person who did not feel contempt at the very idea the words create is not one to brave the world . . . There seems a touch of Common Place vanity in that dread of being obscure & insignificant—you will never be that—& still more *I* am not a person who in any event could give you cause to feel that *I* had made you so. Whatever you may think *I* could never be either of those words.[85]

Harriet Taylor's vanity no doubt transcended the "Common Place," and Mill had offended it mightily. She would have him understand that she knew more than he did and she felt more than he did. No one with a character such as hers could ever be "obscure & insignificant." The love she felt for him was in itself sufficient guarantee that he could not be these things.

Neither party could escape the occasional bouts of dejection thrown up by the intractable problems they faced. The disparity between feelings and action sometimes proved oppressive. Her situation was more difficult than his. India House business, journalism and projects of a political character, a keen engagement with issues central to his work on logic, his developing friendship with Carlyle—these enterprises gave him much to think on that did not involve Harriet Taylor (unless he could not resist pondering the potential harm their relationship could inflict on his drive for public usefulness and distinction). She, on the other hand, lived her daily life within a family whose welfare depended in large measure on her manner of negotiating conflicting personal claims and needs.

A prodigious emotional resilience and a crafty self-fashioning helped carry Harriet Taylor through the thickets. If feeling could not be vented in action, then feeling would have to be privileged over all else (in theory at any rate). She told Mill: "I have always observed where there is strong feeling the interests of feeling are always paramount & it seems to me that personal feeling has more of infinity in it than any other part of character . . . All the qualities on earth never give happiness without personal feeling—personal feeling always gives happiness with or without any other character."[86] Although this conviction did not immunize Harriet Taylor against debility, it did bolster the elevated conception she had of her own individuality and self-worth. Her incandescent devotion to personal feeling, at the same time, did nothing to weaken a well-honed capacity for personal calculation. Harriet Taylor strove to achieve an authoritative influence over the emotional lives of those with whom she was intimate. Mill, much less sure of himself in this realm of experience, was susceptible to this influence.

Harriet Taylor's influence on his emotional life during the 1830s did not spill over into the intellectual and political spheres. In the *Autobiography* he unequivocally stated that by 1830 the "only actual revolution which has ever taken place in my modes of thinking was already complete."[87] Harriet Taylor had no significant impact on his "modes of thinking." When she declared that she knew things of which he was ignorant, she was making no claim for her superior learnedness or intellectual acuity, but rather for her surpassing insight into matters involving "human feeling and character."[88] Mill's remarkable fund of knowledge and singular mental power were certainly an "originating cause" of her interest in him. Harriet Taylor deferred to his philosophical acumen and mastery of the scientific method; she never deluded herself into thinking she could become his intellectual equal. Her initial influence lay in the aesthetic domain, as Mill himself acknowledged in the Early Draft of his autobiography.

> The poetic elements of her character, which were at that time the most ripened, were naturally those which impressed me first, and those years were, in respect of my own development, mainly years of poetic culture. My faculties became more attuned to the beautiful and elevated, in all kinds, and especially in human feeling and character, and more capable of vibrating in unison with it; and I required, in all those in whom I could take interest, a strong taste for elevated and poetic feeling, if not the feeling itself.[89]

A careful reading of this passage reveals Mill's own view of the limited nature of Harriet Taylor's influence in this period.[90] His years of "poetic culture" had begun before he met her, as he made clear when discussing the aftermath of his mental crisis. He was impressed by the "poetic elements" in her character because he was predisposed to value such elements. She noticeably heightened tendencies that had gathered force in the several years leading up to 1830 (this heightening did not constitute a transformation in his "modes" of feeling). She did not decisively alter the course of his development.

CHAPTER FOUR

*Mystifying the Mystic:
Mill and Carlyle in the 1830s*

The intimate relations Mill developed with Harriet Taylor in the early 1830s produced a growth in both the intensity and complexity of his emotional life. Intensity and complexity of a different sort tinged Mill's friendship with Thomas Carlyle in these same years. This friendship encapsulated important elements in Mill's evolving self-definition, and in subtle ways showed the travails attendant upon his pursuit of self-understanding. James Mill, Harriet Taylor, Thomas Carlyle—the force of character exhibited by each of these individuals bespoke a self-confidence that John Mill lacked. He had great powers of mind and a fine and sympathetic spirit, and these qualities made him a person worth knowing. But what did it mean "to know" John Mill? In the early 1830s James Mill would not have presumed to answer this question. Although exasperated at times, Harriet Taylor handled the issue by asserting that she knew John Mill better than he knew himself; what he might have believed he knew about himself counted for less than what she certainly knew about him. Both James Mill and Harriet Taylor had a lot more at stake in their connection with John Mill than did Thomas Carlyle. Yet what Carlyle learned about John Mill in the early 1830s made him want to know more. This desire tested Mill's trust, not of Carlyle but of himself.

★ ★ ★

Thomas Carlyle's personal background paralleled James Mill's in certain respects. Each was born into a Scottish family of small means, and benefited from Scotland's extensive network of parish, village, and grammar schools. Each matriculated at Edinburgh University with a view to entering the ministry, only to abandon this vocational goal. Each also eventually settled in London. When James Mill did so, he had no wish to advertise his humble Scottish origins. Carlyle—in dress, manners, speech, and sentiments—proudly identified with the place, stock, and culture from which he came. He lived over half his life in London. When that life was done, his body, in keeping with his wishes, was buried alongside his parents' graves in the churchyard at Ecclefechan, county Dumfriesshire.[1]

Thomas Carlyle had been born here in December 1795, to hard-working, God-fearing folk. One of nine children, he was reared in a household steeped in a sternly Calvinistic piety. His exceptionally powerful mind made its presence felt early, and his parents were prepared to devote a portion of their meager resources to help this unusually able son equip himself for a career in the ministry. Carlyle's Edinburgh University experience corroded his religious faith, and in 1817 he gave up the idea of becoming a minister. For Carlyle, absence of faith meant not liberation but deprivation. He endured much misery in the several years after 1817, not knowing what to make of his life in a universe devoid of spiritual and moral meaning. An acutely sensitive nervous constitution and frequent bouts of severe indigestion aggravated his existential anguish. He lacked direction and struggled to keep body and mind together while living in cheap lodgings in Edinburgh. The little money he brought in from sporadic work as a tutor and a writer of encyclopedia articles provided no basis for a livelihood. Produce from his parents' farm supplemented what he purchased in the market. The steadfast emotional support given by his family was equally indispensable to Carlyle during these difficult years.

His fortunes received a boost in 1822, when he was introduced by Edward Irving to the Buller family. Irving, whose career as a London preacher would subsequently bring him fame and notoriety, befriended Carlyle when both briefly taught in schools in Kirkcaldy following Carlyle's university years. Charles Buller (the elder), the younger son of a West Country gentry family, had done well in India as a member of the East India Company's Board of Revenue in Bengal. His wife heard a sermon given by Irving in December 1821, and later approached him for help in choosing a tutor for the Bullers' two sons, Charles and Arthur. Irving put in a good word for Carlyle, whose appointment as tutor carried a salary of £200 per annum.[2]

By this time Irving had also introduced Carlyle to Jane Baillie Welsh, a beautiful, brilliant, accomplished, and forthright young woman who lived with her mother in Haddington, a town near Edinburgh. Irving had come to know the family in 1810, when he took up a teaching position at a school recently established in Haddington. John Welsh, a respected and prosperous physician, arranged for Irving to tutor his only child, who had announced that nothing would stand in the way of her learning Latin (she was then nine years old). When Irving introduced Carlyle to his former pupil in 1821, he was himself in love with Jane Welsh (and she with him). Irving, however, was pledged to another, and could not free himself from the commitment he had undertaken. Carlyle, then unaware of these entanglements, embarked on a campaign to make Jane Welsh his wife.[3] Dr. Welsh had died in 1819, and his widow did not consider Carlyle a fit suitor for the hand of her much-sought-after daughter. The aggressive, rustic, and rawboned Carlyle lacked refinement, and his economic prospects seemed pretty dim in 1821. He nonetheless pressed his suit relentlessly, and with time Jane Welsh came to see the remarkable qualities of mind and spirit he possessed. Although not for want of trying, the mother could not hold the daughter back. In 1826 Jane Welsh became Jane Carlyle—a mercurial marriage it would prove to be.

Carlyle left his employment with the Bullers in mid-1824, when Charles Buller the younger enrolled at Trinity College, Cambridge.[4] For a livelihood he now relied upon the writing of periodical articles, the great majority of which dealt with aspects of German literature. In the decade after 1822 Carlyle produced a score of major essays in this field. Having begun to learn the language in 1818–19, Carlyle discovered in German literature a deep reservoir of intellectual and spiritual nourishment. Of the writers associated with the German literary renaissance of the late eighteenth and early nineteenth centuries none mattered more to Carlyle than Goethe. He felt a powerful affinity with Goethe's idea that the most dynamic and complete expression of man's creativity was manifested in a striving to become all that his capacities would allow. The beauty of language, sensitivity of spirit, and force of intellect with which Goethe explored this theme in his works *Faust* and *Wilhelm Meister* moved Carlyle profoundly. In a letter of October 1821 he said of Goethe: "I have an immense love for the man . . . I would travel *above* fifty miles on foot to see Goethe."[5] Carlyle's translation of *Wilhelm Meister*, published in 1824, did much to advance his reputation as an authority on German literature in Scottish and English literary circles. Goethe exemplified the heroic mission of the man of letters, a notion with which Carlyle ardently identified. Carlyle's conception of the heroic individual also owed something to the writings of the German idealist philosopher Johann Fichte. Fichte stated: "The original Divine Idea of any particular point of time remains for the most part unexpressed, until the God-inspired man appears and declares it."[6]

Carlyle believed himself to be a "God-inspired man." His absorption in German literature revived his faith in a spiritual order that transcended the material world. The physical phenomena of the latter offered tangible landmarks that could guide the poet toward the nonmaterial ultimate reality, the Divine Idea. For Carlyle all genuinely creative writing was poetic. The poet sees what others cannot. "He is a *vates*, a seer; a gift of vision has been given him."[7] The poet had a responsibility to transmit his vision of truth to ordinary men and to bring home to them the mandates of the spiritual order. Carlyle sought to make this exalted mission his own.

He perhaps thought of Craigenputtock as the right sort of place for a prophet to dwell. A farm situated in a wild and desolate landscape, Craigenputtock was owned by the Welsh family. The fact that Carlyle had close kin living fairly near Craigenputtock added to its value in his eyes. Having spent the first two years of their married life in Edinburgh, Thomas and Jane Carlyle moved to Craigenputtock in 1828. Over the next few years several commissioned essays written for the *Edinburgh Review* occupied Carlyle, as did the extraordinary manuscript that ultimately became *Sartor Resartus*. As always, he also read voraciously. The farm itself was worked (until 1831) by his brother Alexander. From his Craigenputtock base Thomas Carlyle kept up an active correspondence with another brother, John. In February 1831 he received from this brother, then in London, a letter describing a visit he had just paid Charles and Arthur Buller. The latter had noted that a son of James Mill had written many articles for Albany Fonblanque's *Examiner*. John Carlyle learned that this son was responsible for the series of essays on "the spirit of the age," which had caught Thomas's attention. Arthur Buller depicted John Mill as "a strange enthusiast, with many capabilities but without much constancy of purpose, & without any fixed religion to guide him . . . he has a taste for poetry & can cry as a child over Wordsworth's."[8]

By the time Carlyle got this letter, three of John Mill's "Spirit of the Age" pieces had appeared in the *Examiner*. That they should excite Carlyle's notice is understandable. In June 1829 the *Edinburgh Review* had printed Carlyle's essay titled "Signs of the Times," which had asserted that "great outward changes are in progress . . . The time is sick and out of joint . . . The thinking minds of all nations call for change. There is a deep-lying struggle in the whole fabric of society; a boundless, grinding collision of the New with the Old."[9] The author of "The Spirit of the Age," in his own fashion, spoke to these themes. Carlyle assumed this author to be Fonblanque. The discovery that a son of James Mill was the maker of these essays no doubt quickened his interest. Carlyle scorned James Mill's way of thinking; James Mill, for his part, wanted no truck with Carlyle's "mysticism." In late 1827 Carlyle had sought to obtain a professorship at the new London University. Charles Buller lobbied on behalf of his former tutor and urged James Mill to consider Carlyle's claims. In a letter Carlyle wrote his brother John in March 1828, he caustically noted that Buller "found . . . my German Metaphysics were an unspeakable stone of stumbling to that great Thinker."[10] In connection with the same transaction, Carlyle later conveyed his contempt for the Benthamites. "Nay, is it not true, and clear as day, that I do reckon Jeremiah Bentham no Philosopher, and the Utilitarian system little better than the gross Idol-worship of a generation that has forsaken and knows not the 'Invisible God'?"[11] Carlyle regarded Utilitarianism as a "mechanical" philosophy that made the fundamental error of supposing institutional reform could yield valuable outcomes. He declared: "It is not by Mechanism, but by Religion; not by Self-interest, but by Loyalty, that men are governed and governable."[12] The information that James Mill's son should be "a strange enthusiast" of great ability but no fixed purpose, one who responded to poetry with strong emotion, heightened Carlyle's curiosity about the author of "The Spirit of the Age." He would soon meet the man himself.

★ ★ ★

In August 1831 Carlyle left Craigenputtock for London. He went in search of a publisher for *Sartor* and to see whether London might be a fit place for the Carlyles to reside. Although Craigenputtock had proved an apt venue for Carlyle's regimen of hard concentration and earnest composition, it could not satisfy the longing felt by both husband and wife for companionship beyond the home. A Londoner, William Empson, professor of law at the East India College and a frequent contributor to the *Edinburgh Review*,[13] was known to both Carlyle and John Mill. Empson kept lodgings in the Temple, where Carlyle visited him on 29 August. That night Carlyle wrote his wife a long letter, in which passing reference was made to John Mill. "Of young Mill (the *Spirit of the Age* man) he [Empson] speaks very highly, as of a converted Utilitarian, who is studying German: so we are all to meet, along with a certain Mrs. Austen [Sarah Austin] a young Germanist and mutual intercessor (between Mill and Empson)."[14]

Carlyle and Mill were both guests of the Austins on 2 September (a bad cold prevented Empson from joining the party). The impression Mill made on Carlyle was vividly recorded in a letter the latter wrote Jane Carlyle two days after the gathering. (For the deployment of nonvivid language Carlyle was perpetually at a loss.) John Mill, Carlyle reported, was

> modest, remarkably gifted with precision of utterance; enthusiastic, yet lucid, calm; not a great, yet distinctly a gifted and amiable youth. We had almost four hours of the best talk I have mingled in for long. The youth walked home with me almost to the door; seemed to profess almost as plainly as modesty would allow that he had been converted by the Head of the Mystic School [Carlyle], to whom personally he testified very hearty-looking regard.[15]

On the evening of 13 September Mill visited Carlyle and the two men talked "till near eleven." Mill's rating with Carlyle held fast: "a fine, clear Enthusiast, one who will one day come to something. Yet to nothing Poetical, I think: his fancy is not rich; furthermore he cannot *laugh* with any compass. You [Jane Carlyle] will like Mill."[16]

What did Mill make of Carlyle during the introductory phase of their association? The earliest evidence appears in a letter Mill wrote John Sterling on 20 October. He noted that for some time he had "had a very keen relish" for several of Carlyle's periodical essays, "which I formerly thought to be such consummate nonsense" (this alteration, as Mill understood it, surely being one more sign of his own expanding sympathies). Carlyle in the flesh struck Mill as less "the reflexion or shadow of the great German writers" than he had been disposed to think of him; nonetheless, Carlyle's "mind has derived from their inspiration whatever breath of life is in it." He praised Carlyle for his "liberality & tolerance," which Mill thought "by far the largest & widest . . . that I have met with in any one." Moreover, he appreciated that Carlyle, unlike most critics of the age, "looks for a safe landing *before* and not *behind*: he sees that if we could replace things as they once were, we should only retard the final issue, as we should in all probability go on just as we then did, & arrive again at the very place where we now stand." Mill told Sterling that Carlyle meant to spend the winter in London, which Carlyle wanted to know better. "He is a great hunter-out of acquaintances; he hunted me out, or rather hunted out the author of certain papers in the Examiner." Mill considered the making of Carlyle's acquaintance "the only substantial good" he had gained from writing "The Spirit of the Age."[17] Carlyle rose in Mill's estimation during the winter of 1831–32 (the Carlyles returned to Scotland in

April 1832). In May 1832 he informed Sterling that "Carlyle had passed the whole of a long winter in London; & rose in my opinion, more than I know how to express, from a nearer acquaintance."[18]

Carlyle certainly welcomed this "nearer acquaintance." He valued Mill as an unusually intelligent and amiable companion, a potential disciple worth cultivating, a possible conduit for reaching a wider audience, and a prized practical resource (Mill's large collection of books dealing with the great French Revolution, for example, proved a boon to Carlyle). Reasonably enough, Mill took this interest as a compliment deserving of a gracious response. There were worldly ways in which he could be helpful to Carlyle, and he aimed to be that. He admired Carlyle's deep moral seriousness, his vibrant rhetorical power, his towering creative genius. The distinctive force field Carlyle projected fascinated Mill, whose self-conscious effort to reconfigure himself in the wake of his mental crisis was still very much in progress. For the acquisition of new insights and the assimilation of new truths he had to expose his imaginative faculties to fresh stimuli. At no point would he abandon the disciplined exercise of his critical faculties; to grow, however, he needed to discover what could be gleaned from men of indubitable genius whose way of seeing the world differed fundamentally from his own. In Mill's judgment, Carlyle had the credentials.

The internal jostling of new insights and old truths caused Mill plenty of perplexity (his developing relationship with Harriet Taylor could scarcely act as a countervailing influence). "I found the fabric of my old and taught opinions giving way in many fresh places, and I never allowed it to fall to pieces, but was incessantly occupied in weaving it anew. I never, in the course of my transition, was content to remain, for ever so short a time, confused and unsettled."[19] He did not say that his bewilderment lasted but a short time; only that he was not "content" for it to persist. His correspondence with Carlyle illuminates the problem.

On the last day of July 1832 Carlyle mentioned Mill in a letter he wrote his brother John. "Mill's letters are too speculative," he remarked, "but I reckon him an excellent person, and his love to *me* is great!"[20] Carlyle had shortly before received a letter from Mill in which the latter had distinguished the cultural functions of the artist from those of the logician, and had personalized the distinction.

> I am rather fitted to be a logical expounder than an artist. You I look upon as an artist, and perhaps the only genuine one now living in this country: the highest destiny of all, lies in that direction; for it is the artist alone in whose hands Truth becomes impressive, and a

living principle of action. Yet it is something not inconsiderable (in an age in which the understanding is more cultivated and developed than any of the other faculties, & is the only faculty which men do not habitually distrust) if one could address them through the understanding, & ostensibly with little besides mere logical apparatus, yet in a spirit higher than was ever inspired by mere logic, and in such sort that their understandings shall at least have to be *reconciled* to those truths, which even then will not be *felt* until they shall have been breathed upon by the breath of the artist. For, as far as I have observed, the majority even of those who are capable of receiving Truth into their minds, must have the logical side of it turned *first* towards them; then it must be quite turned round before them, that they may see it to be the same Truth in its poetic that it is in its metaphysical aspect. Now this is what I seem to myself qualified for, if for any thing, or at least capable of qualifying myself for; and it is thus that I may be, and therefore ought to be, not useless as an auxiliary even to you, though I am sensible that I can never give back to you the value of what I receive from you.[21]

Did Carlyle find Mill's discussion of this distinction "too speculative"? Probably not, especially since it included the unmistakably concrete suggestion that the letter's recipient was "perhaps the only genuine" artist "now living in this country." Besides, Mill had used the conceptual discrimination to throw light on the personal. A better candidate for the "too speculative" is a conspicuously opaque meditation that appears earlier in the letter.

Were it not that imperfect and dim light is yet better than total darkness, there would be little encouragement to attempt enlightening either oneself or the world. But the real encouragement is, that he who does the best he can, always does some good, even when in his direct aim he totally fails. For although the task which we undertake is, to speak a certain portion of precious Truth, and instead of speaking any Truth at all, it is possible our light may be nothing but a *feu follet*, and we may leave ourselves and others no wiser than we found them; still, that any one sincere mind, doing all it can to gain insight into a thing, and endeavouring to declare truthfully all it sees, declares *this* (be what it may), is itself a truth; no inconsiderable one; which at least it depends upon ourselves to be fully assured of, and which is often not less, sometimes perhaps more, profitable to the hearer or reader, than much sounder doctrine delivered without intensity of conviction. And this is one eternal and inestimable preeminence (even in the productions of pure Intellect) which the doings of an honest heart possess over those men of the strongest and most cultivated powers of mind when directed to any other end in preference to, or even in conjunction with, Truth. He who paints a thing as he actually saw it, though it were only by an optical illusion, teaches us, if nothing else, at least the nature of Sight, and of *spectra* and phantasms: but if somebody has not seen, or even believed that he saw, anything at all, but has merely thrown together objects and colours at random or to gain some point, it is all false and hollow, and nobody is the wiser or better, or ever can be so, from what has been done, but may be greatly the more ignorant, more confused, and worse.[22]

This rumination was meant to evince "a spirit higher than was ever inspired by mere logic." Instead it showed the disarray occasioned by Mill's "fabric . . . giving way in . . . fresh places," and the execrable effects produced by the improvised introduction of Carlylean thread.

In a letter he wrote Mill in mid-October 1832, Carlyle expressed a keen interest in the second generation of Utilitarians. "Young minds . . . will not *end* where they began: under this point of view, you and certain of yours are of great interest for me."[23] Mill stated in reply that none of the younger Utilitarians had undergone a metamorphosis comparable to his own.

None . . . of them all has become so unlike what he once was as I myself, who originally was the narrowest of them all, having been brought up more exclusively under the influence of a peculiar kind of impressions than any other person ever was. Fortunately however I was not *crammed*; my own thinking faculties were called into strong though partial play; & by their means I have been enabled to *remake* all my opinions.[24]

The ingredients Mill tossed into his remaking-of-opinions blender in the early 1830s sometimes did not easily coalesce.

★ ★ ★

Unsettled though some of his opinions were, Mill had no doubt that his thoughts had a robustness that his feelings lacked. Of this lack he was acutely aware during the first half of 1833, when he found himself wholly unable to plot either the emotional or practical coordinates of his relationship with Harriet Taylor. (Carlyle as yet knew nothing of Mill's ties to Mrs. Taylor.) In March 1833 Mill, responding to Carlyle's wish that he share his feelings as well as his thoughts, declared: "when I give my thoughts, I give the best I have." Carlyle had expostulated on "the boundless capacity Man has of loving." Mill conceded that "in *some* natures" this capacity was indeed "immeasurable and inexhaustible"; not so in others, however. Citing his own case, he could not help but "wonder . . . at the limitedness and even narrowness of that capacity," adding that this constriction "seems to me the only really insuperable calamity in life; the only one which is not conquerable by the power of a strong will." He differed from most people endowed with a feeble capacity for loving in being aware of the deficiency and its import. He was "painfully conscious of that scantiness as a *want* and an imperfection: and being thus conscious I am in a higher, though a less happy, state, than the self-satisfied *many* who have my wants without my power of appreciation."[25]

Mill had given Carlyle good reason to think that the son of James Mill was well on his way to becoming a Carlylean neophyte, evidence for which extended beyond the manner and matter of his letters to Carlyle. Essays Mill published in 1832 and 1833 breathed Carlyle's impact. The October 1832 issue of the *Monthly Repository* included Mill's piece "On Genius," the first paragraph of which invoked an idea forcibly developed in Carlyle's essay "Characteristics" (published by the *Edinburgh Review* in December 1831). Mill declared:

> You judge of man, not by what he does, but by what he is. For, though man is formed for action, and is of no worth further than by virtue of the work which he does; yet (as has been often said, by one of the noblest spirits of our time) the works which most of us are appointed to do on this earth are in themselves little better than trivial and contemptible: the sole thing which is indeed valuable in them, is the *spirit* in which they are done.[26]

The palpable presence of the same noble spirit turned up in Mill's April 1833 *Monthly Repository* essay on the "Writings of Junius Redivivus" (Junius Redivivus was the pseudonym of William Bridges Adams, a frequent contributor to Fox's journal).[27] The following passage, taken from the second paragraph of this essay, is undiluted Carlyle:

> Let the word be what it may, so it be but spoken with a truthful intent, this one thing *must* be interesting in it, that it has been

spoken by man—that it is the authentic record of something which has actually been thought or felt by a human being. Let that be sure, and even though in every other sense the word be false, there is a truth in it greater than that which it affects to communicate: we learn from it to know one human soul. "Man is infinitely precious to man" [words written by Carlyle in a letter to Mill of January 1833], not only because where sympathy is not, what we term *to live* is but to *get through* life, but because in all of us, except here and there a star-like, self-poised nature, which seems to have attained without a struggle the heights to which others must clamber in sore travail and distress, the beginning of all nobleness and strength is the faith that such nobleness and such strength have existed and do exist in others, how few soever and how scattered.[28]

Carlyle wrote Mill on 1 May 1833, a day after seeing a newspaper that printed an extract from this *Monthly Repository* essay. The passage quoted, observed a pleased Carlyle, "was curiously emblematic of my own late thoughts. If it was not you that wrote it . . . then there must be another Mystic in England, whose acquaintance I should gladly make."[29]

In permitting himself the indulgence of experimenting with the part of "Mystic," Mill had raised Carlyle's hopes for a complete conversion. Mill knew such hopes to be illusory, and that he had a duty to tell Carlyle as much. He broached the subject in a letter of 18 May, confessing that he had demonstrated "something like a want of courage in avoiding, or touching only perfunctorily, with you, points on which I thought it likely that we should differ." Mill linked this tendency with his "reaction from the dogmatic disputatiousness of my former narrow and mechanical state" (this would seem to extenuate, if only indirectly, his "want of courage"). The future author of *On Liberty* went on to say that he had no "great notion of the advantage of what the 'free discussion' men, call the 'collision of opinions.'" He had become convinced that "Truth is *sown* and germinates in the mind itself, and is not to be struck *out* suddenly like fire from flint by knocking another hard body against it." Hence he had habituated himself to learning "by inducing others to deliver their thoughts, and to teach by scattering my own," a mode of proceeding generally incompatible with participation in controversy. The rub was that he had pressed his "doctrine and . . . practice much too far," and possibly misled people he esteemed into thinking that he was "considerably *nearer* to agreeing with them" than was in fact the case. In striving to remedy this defect, Mill would in future openly register dissent when he found his views at variance with those of Carlyle, "even though the consequence should be to be lowered in your opinion." Such a result "would shew that for being thought so highly of I had been partly indebted to not being thoroughly known—which I am sure is the case oftener than I like to think of."[30]

Mill must have used much of his courage in saying this much to Carlyle in May 1833; if he had any left, he was not, for the time being, ready to spend it on providing his friend with a detailed list of those matters on which they did not see eye to eye. Rather than move forward to expose areas of disagreement, Mill sought to hold Carlyle's good opinion without specifying significant differences between them. Carlyle consistently preached the immeasurable value of "Silence" to the human soul, and Mill was more than willing to keep his own counsel, while avoiding that of others, at this juncture. "I am the least *helpable* of mortals," he told Carlyle. "I have always found that when I am in difficulty or perplexity of a spiritual kind I must struggle out of it by myself." Mill chose his words with care, and Carlyle could scarcely take exception to the sentiment they expressed. His preferred means of dealing with spiritual distress, Mill confided, was to "shut myself up from the human race, and not see the face of man until I had got firm footing again on some solid basis of conviction, and could turn what comes into me from others into wholesome nutriment." Mill saw no "advantage in communion with others while my own mind was unsettled at its foundations," and he made sure Carlyle comprehended how radically unsteady he felt. "I am often in a state almost of scepticism, and have no theory of Human Life at all, or seem to have conflicting theories, or a theory which does not amount to a Belief."[31] All this was well calculated to elicit Carlyle's sympathy and understanding. Mill did not say he had a "theory of Human Life" different from Carlyle's. Might not he emerge from this bad patch with a firm conviction that Carlyle's "theory of Human Life" was the best one going?

Mill's letter of mid-May had done nothing to diminish Carlyle's estimation of him. If anything, Mill had gained rather than lost ground. In mid-June Carlyle pronounced: "You do well, and needfully, to vindicate your rights of ME-*hood*, having well admitted so many rights of THOU-*hood*." The way Mill had vindicated these rights understandably did little to trouble Carlyle as he pondered the future of their association. Mill's manner of looking at the world could not be expected to correspond exactly with his own; the position from which every man fixes his gaze, "if he be a man at all," must have its own particularities. Provided he "note faithfully and believe heartily what he *sees* there," the result "will not contradict his as faithful brother's view, but in the end complete it and harmonize with it." Carlyle assured Mill that his recent letter had merely confirmed what Carlyle already knew: "that you and I differed over a whole half-universe of things"; he rejected, however, the intimation that "we are moving *from* each other." In Carlyle's judgment, the opposite was the case. In any event, they should not hesitate to bring forward "the friendly conflict of their differences." Each man acknowledged "the infinite nature of truth," the essential foundation "of all profitable communion." Carlyle encouraged his cherished friend "to show me the whole breadth and figure of your dissent." He plainly did not fear the outcome. "About you I will not prophecy here; meanwhile I have my own anticipations, and in any conceivable case must watch you with deep interest."[32]

Mill did not mind either the "deep interest" or Carlyle's reticence respecting "anticipations." In his letter of 5 July, Mill welcomed Carlyle's recognition that "we *still* differ in many of our opinions"—Mill's construction of Carlyle's "whole half-universe" metaphor—and declared that he would not be shy in naming the points on which he thought they disagreed. Yet he continued to be timid. This letter hardly began to reveal "the whole breadth and figure" of his "dissent." He did take the opportunity, however, to say more about the circumstances that had occasioned his letter of mid-May, and to let Carlyle know that his state of mind had improved in the intervening weeks. He noted that for quite some time he had persuaded himself that the differences between them were inconsequential, and that these concerned a few "speculative premises" rather than "practical conclusions." Only when he took stock of his convictions, "after a considerable period of fresh thought and fresh experience," did it dawn on him that they "*materially*" deviated from Carlyle's. This realization had produced the letter of mid-May. Mill now told Carlyle that though he then "wrote as if in a sceptical and unsettled state of mind," his having written in this vein "proved" that he had attained "a more settled state." Instead of getting the detailed account of convictional differences that he might justifiably have expected at this point, Carlyle got a nebulous explanation of how Mill had placed himself on "something like a firm footing." "I can hardly say that I have changed any of my opinions," Mill wrote, "but I seem to myself to *know* more, from increased observation of other people, and increased experience of my own feelings." For reasoned exposition Mill substituted a rhetorical idiom that Carlyle would have a hard time objecting to. The expanded "knowledge of Realities" he had recently acquired had given him "additional ground to build upon." If he nonetheless failed to "raise" his "edifice of Thought to a greater height and so look round and see more of Truth than I could see before," this "must be for want of intellect or for want of will."[33]

Both men knew that Mill did not suffer from "want of intellect." In his letter of July 1833 Mill further developed the theme he had introduced the year before when contrasting his own gifts with those of Carlyle. As a "Poet and Artist," Carlyle had an inherent and intuitive access to "the highest truths." Such truths were self-evident, in the sense that when stated, they immediately gained the mind's assent and required "neither explanation nor proof." The artist's mission was to give vivid expression to truths of this kind. Those members of his audience who themselves possessed an intuitive capacity would seize hold of these truths "in proportion to the impressiveness with which the artist delivers and embodies them." Most people, however, lacked this intuitive endowment, and could not fathom the essence of the artist's insights, which to them seemed "as nothing but dreaming or madness." Enter the logician, who, without being an artist himself, had the intellect and sensibility to grasp the crucial importance of the artist's cognitions. His role was "to supply a logical commentary" on the artist's "intuitive truths" in order "to convince him who never could *know* the intuitive truths, that they are not inconsistent with anything he *does* know; that they are even very *probable*, and that he may have faith in them when higher natures than his own affirm that they are truths." This "humbler part" Mill assigned himself, "a man of speculation" with a reverence for poetry. He had the ability to "feel it and understand it," and the means to "make others who are my inferiors understand it in proportion to the measure of their capacity." Humbler, perhaps, but absolutely vital in Mill's view: "such a person is more wanted than even the poet himself."[34]

Mill's letter of mid-July left Carlyle none the wiser about the "negative part of the relation" between them (Mill's terminology). Whenever he got near a line whose crossing implied a greater opening up—a line toward which his own rectitude impelled him—he withdrew. His personal exchanges in the early 1830s, with Harriet Taylor no less than Thomas Carlyle, did not flow freely (and with his father they scarcely flowed at all). The circumspection evident in the essay he wrote for Taylor on marriage and divorce also turned up in his letters to Carlyle. His struggle to get his thoughts in order had something to do with this tendency, as did his scrupulous regard for the complexity of the issues with which he found himself grappling. All the same, the cultural role he envisaged for himself implied a measure of confidence that these challenges could be surmounted. (It should be noted here that during the early 1830s Mill was writing a considerable chunk of what became his *System of Logic*.)[35] Overcoming his "habitual reserve" was a more serious problem. He lacked faith in his own internal coherence and completeness and he had enough self-awareness to discern this lack.

Both Carlyle and Mill had an interest in preserving the balance of mutual esteem they had created. Carlyle believed that a more open-hearted give-and-take would not threaten this interest. Mill's friendship meant a great deal to him, and he yearned to deepen it. Carlyle, as he told his brother John in early October 1833, considered Mill "one of the purest, worthiest men of this country."[36] Moreover, this most estimable young man acted as Carlyle's London lifeline when the latter returned to Scotland following his sojourn in the metropolis during the winter of 1831–32. Hungry for news of the London literary and political scene, and for gossip about acquaintances he and Mill had in common, Carlyle dined on Mill's letters, and let it be known that his appetite continued unabated ("be more and more diligent in writing; that London and you be still kept in some measure present to me").[37] In addition to the letters, Carlyle received from Mill parcels of books related to Carlyle's literary and historical pursuits. These markedly furthered his work, and Carlyle appreciated Mill's consideration and generosity. Having the sympathy and support of such a talented and admirable young man boosted Carlyle's spirits. For his part, Mill got satisfaction from being of service to a man whose genius merited respect and encouragement. And he valued the good opinion of himself that Carlyle's keen interest conveyed. To Carlyle's pleas for a more resonant communion Mill would have to find an answer.

In a letter of 17 December 1833 Carlyle stepped up the pressure. He had previously asked for news of developments in Paris, where he understood Mill had recently spent some weeks (without knowing the precise nature of the "duty" that had taken him there).[38] Mill obliged in late November, giving a fairly detailed account of certain republican leaders and an update on some Saint-Simonian happenings.[39] One of the things that struck Carlyle in this description was the "strange universal hubbub the French are all making (and most of us make) about the 'good of the species,' and such like." He went on to state emphatically his antipathy to these notions. "How each man seems to mind all men's business,—and leave his own to mind itself! *Something* is to be done; but not for Me or for Thee; no, for Mankind,—when I and Thou are quite past helping." Carlyle felt sure that those who had done most for mankind had sought "guidance and purpose . . . much nearer home"; they had focused on "the working out of what was best and purest" in themselves. "The Good of the Species," he insisted, should be entrusted "to God Almighty the All-Governing," who alone could "comprehend it." This, Carlyle himself certainly intended to do, in the belief "that no good thing I can perform, or make myself capable of performing, *can* be lost to my Brothers, but will prove in reality all and the utmost that I was capable of doing for them." He then directly put the question to Mill: "Now what think you of this Creed, my dear Friend? It is a point which I have long seen we differed in; but seen also, with great pleasure, that we were approximating in. If you still differ from me, even with vehemence, I will not take it ill: in the calmest manner . . . I will appeal to the future John Mill, and he shall decide between us."[40]

Implicit in Carlyle's framing of the question was the idea that Mill—wittingly or otherwise—had withheld what between friends should be freely given: an honest reckoning of belief. Mill caught the implication and saw that his response should attempt to cancel it. He wrote Carlyle on 12 January 1834. Referring to Carlyle's communication of 17 December, Mill declared: "I feel that letter a kind of call upon me to a more complete unfolding to you of my opinions and ways of thinking than I have ever yet made." He wanted to answer that call, and wanted Carlyle to know that his desire to answer it signified "a great change in my character . . . a change, not from any kind of *in*sincerity, but *to* a far higher kind of sincerity than belonged to me before." Mill outlined why he had formerly practiced a lower "kind of sincerity." When he first met Carlyle, Mill observed, he had not yet moved beyond the reactive phase shaped by his rejection of orthodox Utilitarianism. The struggle to free himself from the intolerant sectarianism he had once so stridently espoused had occasioned inner turmoil and misery. The method he chose for combating both the sectarianism and the turmoil carried him to the opposite extreme. He became excessively "catholic and tolerant, & thought one-sidedness almost the one great evil in human affairs." Seeking to extract and assimilate the core truths from the wide range of opinions and beliefs he encountered, he steered away from discussing potentially contentious points with his interlocutors. "I never made strongly prominent my *differences* with any sincere, truth-loving person; but held communion with him through our points of agreement." Had he persisted in this disingenuous mode, "there could have been little worth in me."[41] Only "lately" had he come to see that it was disingenuous, and that corrective action was needed (he did not, however, specify the agents responsible for this recognition). While he had not sought to give others a false impression of himself, neither had he given them the means to form an altogether true one.

Mill's letter of mid-January served as a down payment on what he figured he owed Carlyle in the way of greater forthrightness. He would make a start on "a more complete unfolding," which could not "be all accomplished at once, but must be gradual."[42] It had to be "gradual" if it were to be "all accomplished"; did Mill really think it would ever be "all accomplished"? His use of conditional clauses prompts doubt. Consider the following:

> Whether if you knew me thoroughly I should stand higher, or lower, either in your esteem or in your affection, I know not; in some things you seem to think me *further* from you than I am, in others perhaps I am further from you than you know. On the whole I think if all were told I should stand lower; but there cannot fail, any way, to be much which we shall mutually not only respect but greatly prize in each other; and after all, this, as you & I both know, is altogether of secondary importance; the first being, that we, and all persons and all things, should be truly seen—and as they are.[43]

If "all were told," Carlyle might know him "thoroughly." This fell well short of a commitment to telling all. Mill was never more honest with himself than when rendering self-referential observations in a conditional form.

How much did Mill tell in mid-January? He essentially confined himself to the question put to him regarding the "good of the species," and the "creed" Carlyle had summarily set forth on the subject. The fulcrum of that creed was Carlyle's faith in "God Almighty the All-Governing," and this stood as a cardinal difference between them. To Mill, "the existence of a Creator" was not "a matter of faith, or of intuition; & as a proposition to be proved by evidence, it is but a hypothesis, the proofs of which . . . do not amount to absolute certainty." He deemed the absence of such faith a personal misfortune, but saw no way of making good this fundamental lack. "As this is my condition in spite of the strongest wish to believe, I fear it is hopeless; the unspeakable good it would be to me to have a *faith* like yours . . . I *am* as strongly conscious of when life is a happiness to me, as when it is, what it *has* been for long periods now past by, a burthen." Feeling as he did, he abjured the idea of "propagating" his "uncertainties." These doubts, he went on to say, were not the result of his logical faculty having got the better of some "higher faculty," which might subsequently reassert itself and reverse the provisional triumph won by logic. Mill instead held that "there is wanting something positive in me, which exists in others; whether that something be, as sceptics say, an acquired association, or as you say, a natural faculty." On the issue of the soul's immortality, Mill took a position that would later be termed agnostic. In his view, adequate grounds did not exist for believing either that "it perishes" or that "it survives." (He did not enter into the question of whether the "it" existed, allowing Carlyle to infer from the way the problem was cast that Mill assumed "it" did.) Mill supposed that none of this would come as a surprise to Carlyle. "I am almost sure that you were not much mistaken in the matter, and yet were not quite certain that you knew." Carlyle now could be certain.

The other important difference Mill felt obliged to point out was that he still considered himself a Utilitarian. The caveats he then submitted, however, went a long way toward vitiating this declaration. Apropos of "secondary premises," he stated that he had virtually nothing in common with those who called themselves Utilitarians. He even ventured to say that he was not "a utilitarian at all, unless in quite another sense from what perhaps any one except myself understands by the word." He did not spell out the divergence on secondary principles, and he did not explain the "sense" in which he understood the word. Making this sense clear, he told Carlyle, would itself require "a whole letter." One thing Mill did make clear in his letter of 12 January was that he regarded "the good of the species (or rather of its several units) to be the *ultimate* end" (the "alpha and omega" of his Utilitarianism, Mill parenthetically observed). At the same time, he granted that the effectual forwarding of this end hinged entirely on the strenuous ethos advocated by Carlyle, with "each taking for his exclusive aim the developement of what is best in *himself*." Like Carlyle, Mill maintained that "every human creature has an appointed task to perform which task he is to know & find out for himself; this can only be by discovering in what manner such faculties as he possesses or can acquire may produce most good in the world."[44]

Whatever Carlyle might have thought of him had he told "all," Mill had no cause to fret over the effects of telling what he had (and he very likely understood this to be so). The down payment could be mistaken for payment in full, judging by the gratification it gave Carlyle. The letter, Carlyle reported, "flatters me, and does more: I feel you much closer to me after it. Truly my dear Mill, you are a most punctual, clear, authentic man." Mill's musings relative to the impact his revelations would have on Carlyle's opinion of him made the latter smile. In that smile "lay a greater kindness than it were good to put in words." No demotion in rank for Mill: "No, my friend, you do not stand lower with me; and *I* rather think you would stand higher still, were the whole known." The "Creed" Mill had voiced, Carlyle remarked, "is singularly like my own in most points,—with this difference that you are yet consciously nothing of a Mystic; your very Mysticism (for there is enough of it in you) you have to translate into Logic before you give it place." He counseled "Patience! Patience! Time will do wonders for us."[45]

Mill refers to this exchange in his *Autobiography*, where he states that "for the sake of my own integrity I wrote to him a distinct profession of all those of my opinions which I knew he most disliked; he replied that the chief difference between us was that I 'was as yet consciously nothing of a mystic.'"[46] What Mill said about God in his letter of mid-January 1834 hardly constituted an "opinion." He described a "condition"—the want of "something positive" in himself—and lamented the void. Mill feared this condition was "hopeless"; Carlyle easily persuaded himself that this fear was unwarranted. When Mill turned from God to Utilitarianism, he emphatically distanced himself from the common understanding of what being a Utilitarian meant without telling Carlyle much about the specific composition of his own distinct brand of Utilitarianism. The deliberately fragmentary account he provided, moreover, underscored their convergence of opinion on the means by which the "good of the species" was to be advanced. There was little in Mill's mid-January letter to discourage Carlyle, and discouraged Carlyle was not. Mill's next letter, written in early March, merely reinforced the impression of compatibility: "if I have any *vocation* I think it is exactly this, to translate the mysticism of others into the language of Argument."[47]

Mill no doubt worried about his "integrity," and not without cause. In a letter of mid-April 1833 he had told Carlyle: "I wish you could see something I have written lately about Bentham & Benthamism—but you can't."[48] At this time Mill did not say why he could not let Carlyle see it. The item in question was an essay titled "Remarks on Bentham's Philosophy." It appeared as an appendix to Edward Lytton Bulwer's *England and the English*, a book published in 1833.[49] In the *Autobiography* Mill mentions this "critical account of Bentham's philosophy," which "for the first time put into print" a portion of his negative "estimation of Bentham's doctrines, considered as a complete philosophy."[50] Mill does not say that *he* "put into print" this "estimation"; nor does he inform readers of the *Autobiography* that Bulwer was not at liberty to name the writer of "Remarks on Bentham's Philosophy."

Although the essay lauded Bentham's peerless eminence as a philosopher of law, it sharply criticized his conception of human nature. This conception had given undue prominence to "the selfish principle." In identifying individual motive with the calculation of pleasurable and painful effects, and maintaining that men will put their own selfish interest before the public interest, Bentham had given his ethical system a low moral tone. The author of "Remarks on Bentham's Philosophy" rejected Bentham's assumption that "all our acts are determined by pains and pleasures *in prospect*," and argued that the "pain or pleasure which determines our conduct is as frequently one which *precedes* the moment of action as one which follows it."[51] So repelled are some men by the mere thought of carrying out certain acts that they cannot bear committing them. The pain determining their conduct occurs before rather than after the act in question. Bentham's catalogue of motives had mistakenly omitted "conscience, or the feeling of duty." Although this feeling might be the product of "association," its force could move men to do "right" and abstain from doing "wrong." Bentham's critic declared: "There are, and have been, multitudes, in whom the motive of conscience or moral obligation has been . . . paramount. There is nothing in the constitution of human nature to forbid its being so in all mankind. Until it is so, the race will never enjoy one-tenth part of the happiness which our nature is susceptible of." In conveying the impression that no relation existed between conscience and conduct, Bentham's writings had been, and remained, the source of "very serious evil."[52]

Bentham had died the year before Bulwer's *England and the English* appeared. James Mill, however, was still very much alive. John Mill was not about to avow his authorship of "Remarks on Bentham's Philosophy." Such an avowal would broadcast his apostasy, and bring him into direct conflict with those who subscribed to Utilitarian orthodoxy. He lived and worked with the most influential Utilitarian in England, a man whose wrath he decidedly wished to avoid. Bulwer had asked him to compose a commentary on the character of Bentham's thought and influence, which John Mill understood would be used in connection with Bulwer's book. The precise use that would be made of it he evidently did not know. Mill alluded to the matter in a letter he wrote Carlyle in early August 1833, soon after the publication of *England and the English*. "I told you in one of my letters that I had been writing something about Bentham & his philosophy; it was for Bulwer, at his request, for the purposes of his book: contrary to my expectation at that time, he has printed part of this paper *ipsissimis verbis* as an appendix to his book: so you will see it; but I do not acknowledge it, nor mean to do so."[53] Just over a year later Mill informed another correspondent, John Pringle Nichol, that the appendix on Bentham in Bulwer's book was his. He added: "It is not, and must not be, known to be mine."[54]

John Mill was obviously pleased with his "Remarks on Bentham's Philosophy," and confident that Carlyle and Nichol would think well of what he had written. Yet the idea that James Mill should discover the authorship of this essay made him shudder. John Mill's "habitual reserve" sprang from the acute trepidation he indelibly associated with James Mill's alarming and pervasive presence. The "circumstances" of his childhood, John Mill believed, had "tended to form a character, close and reserved from habit and want of impulse, not from will," a character "destitute of the frank communicativeness which wins and deserves sympathy."[55] His letter to Carlyle of mid-January 1834 spoke of the "great change" his character had recently undergone, "not from any kind of *in*sincerity, but *to* a far higher kind of sincerity than belonged to me before." Mill, perhaps with justice, would draw a distinction between insincerity and evasiveness. The way he parried Carlyle's probing queries and dealt with "Remarks on Bentham's Philosophy" nonetheless raises questions about the content of this "far higher kind of sincerity." Mill had acquired a knack for winning sympathy despite his continued want of candid "communicativeness."

It would not have been odd for Carlyle to think he had had something to do with the "great change" that had come about in Mill's character. Although Mill drew upon his intellectual history to explain why this change was so significant, he did not say what had brought it about. Giving a full account of the latter would have meant taking Carlyle into his confidence regarding Mill's recent "personal" history. Carlyle had learned in early September 1833 that Mill would not be visiting Craigenputtock that autumn. Paris took precedence, the decisive consideration being the "duty" Mill owed "a person to whom of all persons alive I am under the greatest obligations."[56] Mill did not name this person, or explain the nature of his obligations. Around the same time that he informed Carlyle of this development, he wrote Harriet Taylor the letter that impelled her to announce: "I am glad you have said it—I am *happy* that you have—no one with any fineness or beauty of character but must feel compelled to say *all*, to the being they really *love*, or rather with any *permanent* reservation it is *not* love."[57] Mill's "instinct of closeness"[58] had harmed the relationship that mattered to him more than any other. To redress this damage he tried to open up with Harriet Taylor about his problem with opening up. The Paris rendezvous that followed, carried their intimacy to a new level. Mill no doubt believed that the course of his involvement with Harriet Taylor had a decisive bearing on the "far higher kind of sincerity" he felt able to claim for himself in January 1834.

Mill's association with Harriet Taylor produced an instance of this superior form of sincerity in his dealings with Carlyle. In September Carlyle reported to Mill that he had been reading the memoirs of Mme Roland, "a most remarkable woman; one of the clearest, bravest, perhaps as you say *best* of her sex and country; tho' . . . almost rather a man than a woman."[59] Mill challenged Carlyle's linkage of these admirable qualities with maleness. "There was one thing you said of Madame Roland which I did not quite like–it was, that she was almost rather a man than a woman." Mill asked: "*is* there really any distinction between the highest masculine & the highest feminine character?" The women he knew "who possessed the highest measure of what are considered feminine qualities, have combined with them more of the highest *masculine* qualities than I have ever seen in any but one or two men, & those one or two men were also in many respects almost women."[60] Although John Mill had Harriet Taylor in mind when he raised this objection, he did not feel moved to tell Carlyle of her part in his moral improvement. Mill's reticence on this subject notwithstanding, a change in Carlyle's circumstances in late spring 1834 would make known to him "the person to whom of all persons alive" Mill was "under the greatest obligations."

★ ★ ★

During the first half of June 1834 Thomas and Jane Welsh Carlyle moved into 5 Cheyne Row, Chelsea. The distance between the Mills' house in Kensington and Cheyne Row was only three quarters of a mile.[61] In the months after this move, John Mill was a fairly frequent visitor, and the Carlyle-Mill friendship continued to prosper. A letter Carlyle wrote his brother John in October 1834 referred to Mill as "one of the best people I ever saw, and—surprisingly attached to *me*, which is another merit."[62] He could hardly be expected to find Mill's attachment to Harriet Taylor quite so meritorious.

Separate reports furnished by Sarah Austin and Charles Buller, in May 1834, first alerted Carlyle to Mill's involvement with Harriet Taylor. For Buller it was a subject fit for jest (Buller took an expansive view of what was eligible for such treatment); for Mrs. Austin a matter of alarm. Carlyle tended to the view that Sarah Austin exaggerated when she spoke of Mill "being lost to all his friends and to himself, and what not." He observed: "I traced *nothing* of this in poor Mill; and even incline to think that what truth there is or was in his adventure may have done him *good*."[63] (Carlyle did not then know that Mill's entanglement with Mrs. Taylor predated the inception of his own friendship with Mill—he was certainly in no position to assess her impact.) If anything, this story further piqued Carlyle's already deep interest in "poor Mill."

With the Carlyles settled in London, Mill decided they had better meet Harriet Taylor. He knew they would hear about her from others, and that his failure to break the ice himself could have a chilling effect on his friendship with Carlyle. Besides, he was proud of Harriet Taylor and of her love for him. Mill arranged for her to visit the Carlyles on 21 July 1834, and she did not disappoint. This visit prompted Carlyle's depiction of Harriet Taylor as "a living romance-heroine, of the clearest insight, of the royallest volition."[64] The glow had not faded a fortnight later, though Carlyle suspected the entrancement might prove ephemeral. "We have made . . . a most promising new acquaintance, of a Mrs Taylor," Carlyle wrote his mother in early August. He styled her "a young beautiful reader of mine and 'dearest friend' of Mill's, who for the present seems 'all that is noble' and what not. We shall see how that wears."[65]

Wear well it did not. On 12 August the Carlyles were part of a dinner party hosted by John and Harriet Taylor. The other guests were Mill and W.J. Fox. The previous week, in the letter to his mother, Carlyle had expressed a particular interest in meeting Fox.[66] Writing to his brother John several days after the dinner, Carlyle described Fox as "a little thickset bushy-locked man of five-and-forty, with bright sympathetic-thoughtful eyes . . .[and] a tendency to pot-belly and *snuffiness*." Carlyle said that he would not mind meeting "the man again . . . he professed to be unwell (as I too was), and rather 'sang small.'" (Singing small in Carlyle's presence was not a bad idea; when it came to singing large, no one could surpass Carlyle himself, as least in his own estimation.) The advantage of Harriet Taylor's further company now struck Carlyle as questionable. "Mrs Taylor herself did not yield unmixed satisfaction, I think, or receive it: she affects, with a kind of Sultana noblemindedness a certain girlish petulance, and felt that it did not wholly prosper."[67]

By mid-autumn 1834 Carlyle concluded that Mill had gotten himself mixed up with people whose characters could not withstand scrutiny, close or otherwise. Although his affection for his lovesick friend held fast, Carlyle became increasingly contemptuous of Harriet Taylor and her *Monthly Repository* circle. The row within South Place Chapel produced by Fox's peculiar domestic situation took a public turn in August and September, when Fox offered to resign as minister, received a vote of confidence, and watched his most truculent opponents secede from the congregation. The light thrown on Fox's personal conduct by his wife's offensive had given a special twist to the unorthodox opinions on marriage he had voiced in the *Monthly Repository*. Carlyle's Old Testament sensibilities could not countenance a shirking of moral responsibility in favor of doing what one liked. Before the end of October he had decided that Fox's set had gone wild for just this kind of shirking. That Mill should be "greatly occupied of late" with such people annoyed Carlyle, who complained to his brother John that he "seldom" saw his favorite London friend. (During the initial months of the Carlyles' London domicile Mill had tried to be especially attentive; as they increased their range of contacts, he reduced the frequency of his visits. While he valued Carlyle's company, he prized more highly that of Harriet Taylor.) Carlyle had come to share some of the worry evinced by the Austins.

> It is that fairest Mrs Taylor you [John Carlyle] have heard of; with whom, under her husband's very eyes, he is (Platonically) over head and ears in love! Round her come Fox the Socinian, and a flight of really wretched-looking "friends of the species," who (in writing and deed) struggle not in favour of Duty being *done*, but against Duty of any sort almost being *required* . . . Most of these people are very indignant at marriage and the like; and frequently indeed are obliged to divorce their own wives, or be divorced: for tho' the *world* is already blooming (or is one day to do it) in everlasting "happiness of the greatest number," these people's own *houses* (I always find) are little Hells of improvidence, discord, unreason. Mill is far above all that, and I think will not sink in it; however, I do wish him fairly from it, and tho' I cannot speak of it directly would *fain* help him out.[68]

In this matter, Mill was beyond help, be it Carlyle's or anyone else's. Yet Carlyle's wish to be of assistance attested to the strength of his affection for his friend, who continued to elicit Carlyle's warm sympathy. Carlyle had a gift for friendship, but he did not distribute it widely; he cared for Mill as he cared for few others in the 1830s. The more he learned of Harriet Taylor and W.J. Fox, the more critical of them he became. Yet even Mill's imprudent association with these wayward "friends of the species" did not lessen Carlyle's tender feelings for him. These feelings would be severely tested in March 1835, when Mill inadvertently inflicted heavy damage on the residents of 5 Cheyne Row.

★ ★ ★

Mill and Carlyle shared a passionate interest in the history of the great French Revolution. In the early 1820s, when Mill began serious study of the subject, the history of the Revolution "took an immense hold of" his "feelings."[69] This enthusiasm, once kindled, never waned. "Commenced by the people, carried on by the people, defended by the people with a heroism and self-devotion unexampled in any other period of modern history," the Revolution, for Mill, possessed a singular grandeur.[70] His countrymen's ignorance of the subject appalled him. For most Englishmen, including those "who read and think," the French Revolution conjured up "a dim but horrible vision of mobs, and massacres, and revolutionary tribunals, and guillotines, and fishwomen, and heads carried on pikes, and *noyades*, and *fusillades*, and one Robespierre, a most sanguinary monster."[71] Mill tried to repair some of this ignorance in his *Westminster Review* articles "Mignet's French Revolution" (1826) and "Scott's Life of Napoleon" (1828). The latter, an essay of more than twenty thousand words that trenchantly exposed Sir Walter Scott's misrepresentations of the French Revolution, vigorously defended the moderation of the Constituent Assembly and celebrated Mill's beloved Girondists as "the purest and most disinterested body of men, considered as a party, who ever figured in history."[72] The lengthy essay on Scott made use of the sizeable collection of materials related to the history of the French Revolution that Mill had amassed during the 1820s. Mill would put these materials at Carlyle's disposal.

By the early 1830s Carlyle had come to believe that history offered the key to comprehending divine revelation.[73] History recorded the judgments imposed on past societies that had trespassed against the laws of God, and men in the present should heed the lessons manifested in that record. Carlyle conceived a lofty and prophetic role for the reverential and inspired historian, who would be poet as well as prophet. Carlyle had no particular regard for the writing of verse. When imbued with the right spirit and rendered with great imaginative force, prose works of history embodied creative power of the highest order. Only such power could seize upon concrete elements of the human experience to make known the essence of the Ideal, the true function of Poetry. "Is not all Poetry," he asked Mill rhetorically, "the essence of Reality . . . and true History the only possible Epic?"[74]

Carlyle deemed the French Revolution the perfect subject for the historian who aspired to be both poet and prophet. Writing the history of that Revolution, he declared to Mill in September 1833, "is properly the grand *work* of our era." The "right understanding" of this history, he believed, could yield "all possible knowledge important for us; and yet at the present hour our ignorance of it in England is probably as bad as total." Here was a subject equal to Carlyle's soaring ambitions. "To me," he told Mill, "it often seems, as if the right *History* . . . of the French Revolution were the grand Poem of our Time; as if the man who *could* write the *truth* of that, were worth all other writers and singers. If I were spared alive myself, and had means, why not I too prepare the way for such a thing?"[75]

Carlyle wrote "I too" because he had, in the same letter, urged Mill to tackle the subject in earnest. In his reply, dated 5 October 1833, Mill admitted that the thought of doing so had more than once crossed his mind. If Carlyle himself decided not to take up the French Revolution, wrote Mill, "it is highly probable I shall do it sometime." Why wait? The time was not ripe for giving public voice to thoughts that should be stated in connection with any "true" history of the French Revolution. Mill could not write that history, "so as to be read in England, until the time comes when one can speak of Christianity as it may be spoken of in France; as by far the greatest and best thing which has existed on this globe, but which is gone, never to return, only what was best in it to reappear in another and still higher form, some time (heaven knows when)." The man unprepared to speak out his "whole belief on that point" could not "write about the French Revolution in any way professing to tell the *whole* truth."[76] In principle, Mill liked the idea of speaking his "whole belief." Yet he feared that doing so in defiance of the ruling prejudices of those he hoped to influence could hinder his public usefulness. Then, too, there was the habitual guardedness that made it hard for him to act without first asking himself whether what he wished to say could be "prudently . . . avowed to the world."[77] Unwilling to say all that he wished to say about a subject so important as the French Revolution, he would refrain from purporting to say all while knowing he had omitted something fundamental. If Carlyle was ready to take the field, Mill would leave it to him.

By the summer of 1834 Carlyle was ready. In July he told his mother that he was engaged in preparing himself for the challenge: "I have got a heap of Books about me and am actually employing myself daily in preparation for that Book of my own! It is on the French Revolution, which seems far the eligibilist for my first: there is an appetite for it; there are plenty of Documents and materials; Mill himself laid me out the other day a whole barrowful."[78] Carlyle immersed himself in these documents and materials, and started writing in September. By January 1835 he had finished the first of the three volumes he figured he would need to get the job done. His wife and Mill—"no other knows of it," he informed his brother John in mid-January—had read portions of what he had written, and Carlyle took heart from their response. "Jane says it will do; and Mill."[79] In mid-February Jane Carlyle wrote: "Car[l]yles' [sic] book gives great satisfaction so far as it is gone to John Mill and me."[80] During that month Carlyle put in Mill's hands the manuscript of the entire first volume.

On the evening of 6 March a "rap was heard at the door" of 5 Cheyne Row. Mill

> entered pale, unable to speak; gasped out to my wife to go down and speak with Mrs Taylor [who waited in a carriage outside]; and came forward (led by my hand, and astonished looks) the very picture of desperation. After various inarticulate and articulate utterances to merely the same effect, he informs me that my *First Volume* (left out by him in too careless a manner, after or while reading it) was, except for four or five bits of leaves, *irrevocably* ANNIHILATED! I remembered and can still [the day after] remember less of it than anything I ever wrote with such toil . . . *It is gone*; and will not return. Mill very injudiciously staid with us till late; and I had to make an effort and speak as if indifferently about other common matters; he left us however in a *relapsed* state; one of the pitiablest.[81]

Mill was in no fit state to give a detailed explanation of how such a thing could have happened. Supplying his brother John with a report of the disaster, Carlyle noted that "Mill had left it out (too carelessly); it had been taken for wastepaper: and so five months of as tough labour as I could remember of, were as good as vanished, gone like a whiff of smoke."[82] Manuscripts of one sort or another must have routinely shown up in the Mill household. In ordinary circumstances, John Mill—like his father, a man of meticulous habits—would not have treated any manuscript carelessly, be it his own or one that had been entrusted to him. Radical distractions, however, can cause even the most fastidious of people to act negligently. Mill might have been upset by something that had occurred at home, or by a disturbance in his relations with Harriet Taylor, and insensibly put Carlyle's manuscript in harm's way. Brusque treatment of a manuscript by any member of the Mill household (maids included) surely would have constituted aberrant conduct. Aberrant acts happen.

But did Carlyle's manuscript remain in the Kensington residence occupied by James Mill and his family from the moment John Mill brought it home until its time of destruction? In a letter he wrote Carlyle several days after the terrible evening of 6 March, Mill admitted that he had read aloud to Harriet Taylor "much" of the manuscript that had been lost.[83] Mrs. Taylor did not visit the household of James Mill. If John Mill read to Harriet Taylor a large portion of the manuscript, he must have done so while visiting her. The Taylors' Kent Terrace home seems the likely venue, but another possibility exists. At some point during the second half of the 1830s, John Taylor allowed his wife to take a cottage in the country, first (so far as can be known) at Keston Heath, located near Bromley in Kent, and then at Walton-on-Thames. An August 1837 letter addressed to Harriet Taylor by her sister Caroline shows the Keston Heath address;[84] Carlyle refers to Keston in a letter he wrote Mill in March 1838.[85] Although no earlier mention of Keston Heath appears in surviving correspondence, the idea that Mill could have visited Harriet Taylor at a residence other than Kent Terrace is not outlandish. No certainty can be had regarding the site where the original Volume One of Carlyle's *French Revolution* lay in ashes. Servants employed by the Taylors might have been less careful about manuscripts than servants of the Mills. Such a supposition, however, amounts to very little when it comes to trying to piece together what happened to this particular manuscript.

Years after the event Carlyle persuaded himself that Harriet Taylor bore most responsibility for the loss. He mentioned the episode shortly after being informed of Mill's death by Charles Eliot Norton, telling Norton that the manuscript had been mistakenly burnt by a housemaid in Harriet Taylor's residence. Norton reported Carlyle saying that "She had it at her house on the riverside at Kingston [Carlyle may well have said "Keston"], and I shall never forget the dismay on John Mill's face when he came to tell me that the housemaid had lighted the fire with it."[86] An obituary on Mill in the *Daily Telegraph* (10 May 1873) also linked the misadventure to Harriet Taylor (presumably Carlyle himself, or someone with whom he had shared the story, was the anonymous author's source).[87]

This account irked Mill's sister Harriet, and prompted her to write a letter to Carlyle in mid-May 1873. She could not credit the notion that her brother "would have shewn your manuscript to anyone without your permission"—Carlyle had been given reason to think otherwise—and added: "As far as my recollection goes, the misfortune arose from my brother's own inadvertence, in having given your papers amongst waste paper for kitchen use. I can, perfectly well, remember our search, and my dear brother's extreme distress." Carlyle's reply, dated 17 May 1873, remarked that "in fact my impression really was, that night when your Brother came to us pale and agitated, as I have seldom seen any mortal, that Mrs. Taylor's house and some trifling neglect there, had been the cause of the catastrophe." Carlyle most probably created this impression himself, long after the episode occurred. He did remember, when writing to Mill's sister, that her brother was in such misery on the evening in question that "we had to forbear all questioning on the subject."[88] The relevant documents from March 1835 (especially Carlyle's journal entry of 7 March and subsequent letter to his brother) do not support the contention that he then believed "some trifling neglect" at Mrs. Taylor's home had caused the calamity.

Carlyle's death in early February 1881 triggered further exchanges on the subject. The obituary in *The Times* asserted that Mill had passed the manuscript on to "his future wife. What became of it was never exactly known."[89] A letter to the editor, signed F.W.R., claimed that the manuscript had been done away with by "Mr. Mill's cook." Once again Mill's sister Harriet weighed in, but this time her account clashed with what she had written to Carlyle in 1873. Her letter to *The Times* (17 February 1881) insisted that "Valuable papers were not left about in Mr. Mill's house, nor was the disaster owing to him, beyond the fact that he had lent the manuscript to the person at whose house it was destroyed."[90]

In the end, therefore, Mill's sister upheld Carlyle's version. What both ultimately came to believe about what had happened, however, casts no light on what actually occurred in the early part of 1835. Neither Carlyle nor Mill's sister had reason to remember Harriet Taylor Mill fondly; at one time, they had both felt a deep affection for John Mill, warm memories of whom persisted years after his death. Had Harriet Taylor been involved in some way with the loss of Carlyle's manuscript, Mill surely would not have divulged to Carlyle that he had read sections of it to her. In the same letter that he revealed this, Mill responded to Carlyle's offer to let him read the first book of Volume Two, then nearing completion. Inasmuch as his own trustworthiness had been compromised, he instructed Carlyle to "give it to Mrs. Taylor—in her custody no harm could come to it."[91] Mill would no doubt have gone to great lengths to protect Harriet Taylor. The content of this 10 March letter—especially the disclosure that she had had some contact with the ill-starred manuscript—strongly suggests that he did not think she stood in need of protection in this instance. A fateful and peculiar mix of factors lay behind the disaster. Precisely what figured in that mix remains hidden, and the chain of events producing one of the most dramatic and traumatic episodes in Mill's life cannot be reconstructed.

Both Carlyle and Mill acted admirably in the immediate aftermath of the calamity. Despite the severe and unforeseen blow he had suffered, Carlyle mustered abundant sympathy for the plight of his disconsolate friend, to whom he wrote the day after Mill's painful visit. "How are you? You left me last night with a look which I shall not soon forget. Is there anything that I could do or suffer or say to alleviate you? For I feel that your sorrow must be far sharper than mine; yours bound to be a *passive* one." Carlyle assured Mill that though he could not recreate what had been lost, he could create something anew to stand in its place. Moreover, he had confidence in his ability to complete what he had started. "That I *can* write a Book on the French Revolution is (God be thanked for it) as clear to me as ever; also that, if life be given me so long, I will. To it again, therefore!"[92] For his part, Mill implored Carlyle to let him provide monetary compensation for the time and effort expended in building the manuscript Mill had not kept safe. "I beg of you with an earnestness with which perhaps I may never again have need to ask anything as long as we live, that you will permit me to do this little as it is, towards remedying the consequences of my fault & lightening my self-reproach."[93] Carlyle agreed. On 16 March Mill sent Carlyle a draft for £200 (one-third of John Mill's annual salary); the following day, Carlyle returned this draft to Mill, explaining that the money spent during the writing of the manuscript had not exceeded £100.[94] Mill pressed Carlyle to accept the original figure; and if he would refuse £200, he should at least take £150.[95] Carlyle would not take a penny more than £100, and Mill had to give way.

In 1837 *The French Revolution* was published in three volumes. Mill's review, one that sought to set the tone for the book's reception, appeared in the July issue of the *London and Westminster Review*. Its opening paragraph declared: "This is not so much a history, as an epic poem; and notwithstanding, or even in consequence of this, the truest of histories. It is the history of the French Revolution, and the poetry of it, both in one; and on the whole no work of greater genius, either historical or poetical, has been produced in this country for many years."[96] The laudatory review that followed brought to a close Mill's remarkable personal association with a book that is generally regarded as one of Carlyle's most brilliant literary achievements and as a masterpiece of Romantic history writing.

★ ★ ★

Mill and Carlyle remained on good terms for several years after the publication of *The French Revolution*. After 1840 Carlyle's contempt for democracy and exaltation of power as embodied in the heroic God-inspired individual became evermore pronounced. Mill recoiled from such attitudes, and friendly communication between the two men virtually ceased.

In a practical sense, the Mill-Carlyle friendship during the 1830s mattered more to Carlyle than to Mill. Carlyle was struggling to make a literary reputation for himself, and Mill did his best to help. In the early 1830s Carlyle had failed to get *Sartor Resartus* published as a book, and its publication in installments by *Fraser's Magazine* in 1833 did little to advance his aspirations. Widespread recognition arrived with the publication and reception of *The French Revolution*. The auxiliary support Mill gave this project contributed significantly to its completion. Against this, of course, must be set the manuscript disaster that carried a mental and emotional cost impossible to calculate. Would Carlyle's task have been harder or easier had Mill been uninvolved? Probably harder. Mill's private collection of materials pertinent to the history of the French Revolution was unusually extensive. He lent these materials to Carlyle, who had access to them for so long as needed. Mill also shared his knowledge of France and French history with Carlyle and offered valuable ideas and encouragement. Mill's belief in Carlyle no doubt counted for much less than did Carlyle's belief in himself, but count it did. Carlyle's warm feelings for the friendship he enjoyed with Mill in these years positively affected the spirit in which he tackled the writing of his book.[97] Although Mill's £100 could not bring back Carlyle's lost manuscript, it surely eased some of his monetary worries in the months after March 1835. Carlyle's net gain from Mill's involvement was considerable.

Mill's review of *The French Revolution* figured in that net gain. It praised the book's striking originality, epic breadth of vision, vivid and spellbinding depiction of people and events, and notable freedom from political prejudice. Moreover, the review quoted lengthy passages from Carlyle's text to illustrate its manifold virtues.[98] Carlyle expressed his appreciation: "No man, I think, need wish to be better reviewed. You have said openly of my poor Book what I durst not myself dream of it, but should have liked to dream had I dared."[99] Mill's review helped gain acceptance for a work whose form and style struck some as wildly unorthodox. The measure of critical acclaim won by the book established its author as a major literary figure, an end Mill had striven to promote. In a letter of 1840 he stated that he had "greatly accelerated the success" of Carlyle's book, one "so strange & incomprehensible to the greater part of the public, that whether it should succeed or fail seemed to depend upon the turn of a die—but I got the first word, blew the trumpet before it at its first coming out & by claiming for it the honours of the highest genius frightened the small fry of critics from pronouncing a hasty condemnation, got fair play for it & then its success was sure."[100] His contribution may have been less decisive than he wished to think, but it assuredly helped get "fair play" for Carlyle's extraordinary treatment of the French Revolution.

Upon the substance of Mill's thought and beliefs Carlyle had small impact. The adult son of James Mill did not need to be taught the value of work and the importance of duty. Nor did the stress Carlyle placed on individual moral development seem unfamiliar. Virtue meant as much to James Mill as it meant to Carlyle, and Carlyle's ideal of moral improvement was no more expansive than James Mill's. If the elder Mill's "moral convictions . . . were very much of the character of those of the Greek philosophers," Carlyle's closely resembled those of the Hebrew prophets. What John Mill said of his father relative to his moral convictions—they "were delivered with the force and decision which characterized all that came from him"—could with equal justice be said of Carlyle's delivery.[101] What of Mill's fancy that his vocation lay in translating "the mysticism of others into the language of Argument"? The language of Millian argument could not accommodate Carlyle's mysticism, whose foundations were antithetical to Mill's mode of comprehending "reality." How might Mill have gone about translating into "Argument" the following emblematic passage from Carlyle's *Sartor Resartus*?

> Thus, like some wild-flaming, wild-thundering train of Heaven's Artillery, does this mysterious MANKIND thunder and flame, in long-drawn, quick-succeeding grandeur, through the unknown Deep. Thus, like a God-created, fire-breathing Spirit-host, we

emerge from the Inane; haste stormfully across the astonished Earth; then plunge again into the Inane. Earth's mountains are levelled, and her seas filled up, in our passage; can the Earth, which is but dead and a vision, resist Spirits which have reality and are alive? On the hardest adamant some footprint of us is stamped in; the last Rear of the host will read traces of the earliest Van. But whence?—O Heaven, whither? Sense knows not; Faith knows not; only that it is through Mystery to Mystery, from God and to God.[102]

The Carlylean sentiments Mill voiced in the 1830s were detachable from Carlyle's mysticism, the essence of which could not be rendered in the idiom of Millian argument.

It was the power of Carlyle's creative imagination that resonated with the younger Mill, who took an especially keen interest during the 1830s in the ability of the artist to transmit moral lessons through the vivid and compelling representation of human action and feeling. (This interest is evident in a number of periodical essays Mill wrote in this decade, including "On Genius" [1832], "Thoughts on Poetry and Its Varieties" [1833], "Tennyson's Poems" [1835], "Writings of Alfred de Vigny" [1838], and "Milnes's Poems" [1838].) The originality and power of Carlyle's rhetorical genius colored some of John Mill's own prose during the 1830s, a fact he came to regret. Writing to George Henry Lewes toward the close of 1840, Mill referred to the essay "On Genius" (1832), which Lewes had recently discovered. "The 'Genius' paper is no favorite with me, especially in its boyish stile. It was written in the height of my Carlylism, a vice of style which I have since carefully striven to correct . . . I think Carlyle's costume should be left to Carlyle whom alone it becomes."[103]

In his *Autobiography* Mill declared that "the good" done him by Carlyle's writings "was not as philosophy to instruct, but as poetry to animate." He also observed that he did not consider himself "a competent judge of Carlyle. I felt that he was a poet, and that I was not; that he was a man of intuition, which I was not; and that as such, he not only saw many things long before me, which I could only, when they were pointed out to me, hobble after and prove, but that it was highly probable he could see many things which were not visible to me even after they were pointed out."[104] Mill's "years of poetic culture"[105] overlapped the early phases of his association with Harriet Taylor and his period of close friendship with Carlyle. These years, however, had begun before he met either of them. In different ways, Taylor and Carlyle invigorated

Mill's already awakened appetite for "poetic feeling." Although he could not be a poet himself, Mill needed to believe that he could respond with feeling to the creative genius of others. The notable respect and ardent affection he got from Carlyle during the 1830s helped answer this need.

CHAPTER FIVE

Mill and the Secret Ballot

J.S. Mill's transformation from a fervent advocate of the ballot to a decided opponent of secret voting, although frequently noted, has never been adequately explained. Mill himself provides a simple enough—rather too simple, one suspects—explanation of this change of opinion in *Thoughts on Parliamentary Reform*, which offers his fullest treatment of the question (written in the mid-1850s, this pamphlet was withheld from publication until 1859).[1] There he says that the two decades following the passing of the 1832 Reform Act had fundamentally altered the social and political condition of the country. The state of English society in the 1830s had made a call for the ballot altogether fitting. By the 1850s the adoption of secret voting had been rendered both unnecessary and undesirable.[2] Before considering the content and implications of Mill's argument, the evolution of his position on the issue must be traced. The significance of his ultimate rejection of the ballot can be appreciated only against the backdrop of his avid support of the measure in the 1830s.

In the 1830s the secret ballot numbered Mill among its most devoted and enthusiastic promoters. Joseph Hamburger's valuable work on Mill and the Philosophic Radicals makes clear the importance they attached to the issue.[3] The theoretical foundation for their commitment to the ballot had been shaped by James Mill in his *History of British India* and in an article written for the *Westminster Review* in 1830. He used his discussion of the constitution of the East India Company to set forth his view of the criteria that should determine whether or not the secret ballot ought to be employed.

> A voter may be considered as subject to the operation of two sets of interests: the one, interests arising out of the good or evil for

which he is dependent upon the will of other men: the other, interests in respect to which he cannot be considered as dependent upon any determinate man or men . . . In all cases . . . in which the independent interests of the voter, those which in propriety of language may be called his own interests, would dictate the good and useful vote; but in which cases, at the same time, he is liable to be acted upon in the way of either good or evil, by men whose interests would dictate a base and mischievous vote, the ballot is a great and invaluable security. In this set of cases is included, the important instance of the votes of the people for representatives in the legislative assembly of a nation . . . There is, however, another set of cases, in which those interests of the voter, which have their origin primarily in himself, and not in other men, draw in the hurtful direction; and in which he is not liable to be operated upon by any other interests of other men than those which each possesses in common with the rest of the community. If allowed, in this set of cases, to vote in secret, he will be sure to vote as the sinister interest impels. If forced to vote in public, he will be subject to all the restraint, which the eye of the community, fixed upon his virtue or knavery, is calculated to produce: and in such cases, the ballot is only an encouragement to evil.[4]

These theoretical distinctions bolstered the case for the ballot put forward by the Philosophic Radicals in the 1830s, and J.S. Mill would again draw upon them two decades later when arguing the case against secret voting at parliamentary elections.[5]

James Mill gave detailed application to these principles in a lengthy article on the ballot that appeared in the July 1830 issue of the *Westminster Review*.[6] He argued that secret voting would eliminate intimidation and bribery (illegitimate practices whose efficacy depended on knowing for whom an elector voted), vastly reduce the cost of elections, and ensure that candidates would be judged according to their personal fitness for the duties of legislation rather than by the depth of their purse or the amount of immoral influence they could wield for electoral purposes. James Mill insisted that the legitimate moral influence of property would not be endangered by secret voting. On the contrary, by removing the corrupt influence of property the ballot would "give full scope to the exercise of the moral influence."[7]

One would search in vain for any sort of elaborate theoretical justification for the adoption of secret voting in J.S. Mill's abundant political writings of the 1830s; his support for the ballot, nonetheless, followed from his ideological conception of political forces that sought to obscure the divergent interests of the governing class and the bulk of the population. The Philosophic Radicals, Mill among them, saw the party conflict between Whigs and Tories as factitious. In their view (as Hamburger has so ably shown) both Whig and Tory parties, dominated by the landed interests, shared a fundamental commitment to preserving the political privileges of the aristocracy. The political ascendancy they enjoyed assured that their selfish interests would be promoted at the expense of the rest of the community. According to the Philosophic Radicals, the political division that really mattered was not between Whig and Tory but between the aristocracy and the people. The political activity of the Philosophic Radicals during the 1830s aimed to demonstrate the fallacious nature of the existing party structure and to reshape political alignments to conform to social reality. The genuinely liberal section of the Whig-Liberal party, they held, should join with the radicals, and thereby compel the conservative Whigs to declare openly their allegiance to aristocratic interests. The result would be a struggle between a Radical-Liberal party, representing the interests of the people, and the party of the aristocracy, dedicated to the preservation of aristocratic privilege. The victory of the former would then be only a matter of time.[8]

The Philosophic Radicals regarded the ballot as a vital weapon in this political combat.[9] They contended that intimidation and bribery at elections, which secret voting would neutralize, were widely and effectively employed by the landed interest. The adoption of the ballot would extinguish this pernicious electoral influence and thereby undermine the political dominance of the aristocracy. Additionally, the Philosophic Radicals correctly calculated that the Whig leadership would join the Tories in resisting the enactment of secret voting. The selfish interests they held in common would thereby stand blatantly revealed, as would the counterfeit nature of the party political conflict they had colluded in perpetuating. Hence the importance of the ballot in the eyes of the Philosophic Radicals.

J.S. Mill's numerous articles on contemporary politics in the 1830s persistently attempted to drive home the centrality of the ballot question. These articles are by Mill the political sectarian, not Mill the political philosopher. In his more speculative articles on politics during this period the secret ballot is seldom mentioned. His two reviews of Tocqueville's *Democracy in America* (1835 and 1840) do not touch on the issue; nor does his article "Civilization" (1836).[10] It is true that his 1835 review of Samuel Bailey's *Rationale of Political Representation* does quote a long passage from Bailey strongly favorable to the ballot.[11] This essay also contains the observation, made almost in passing, that "A generation at least must elapse, before an aristocracy will consent to seek by fair means the power they have been used to exercise by foul."[12] When a generation had elapsed, Mill himself decided that the ballot was no longer needed.

Mill's advocacy of secret voting in the 1830s was not about political principle. The value of the ballot stemmed from its place in the struggle to create a viable radical party and to thwart aristocratic political influence. That Mill saw the issue in this light is incontrovertible. As de facto editor of the *London and Westminster Review*, in the latter half of the decade, Mill worked hard to foster the radical cause. The ballot was perhaps the most visible issue in the campaign he conducted. As early as July 1835 Mill made clear to readers of the *Westminster* the critical importance of the ballot. The general election held at the beginning of this year significantly enlarged the Tory presence in the House of Commons, though not by a magnitude sufficient to keep Peel's ministry in office. Melbourne's second administration took office in April. Mill spelled out his view of the problem facing the Whig government:

> without the ballot we shall speedily have a Tory parliament;[13] and ... the present ministers will have to decide, whether they will support the ballot, or abandon office to the Tories, or coalesce with the Tories on their own terms. The exact time when this decision must be made it is impossible to foresee, but by no power can it be postponed for more than a year or two. When it comes, which course will the ministers choose? Probably they will not all of them make the same choice. The problem will then be reduced to its simplest terms: Who is for the aristocracy and who for the people, will be the plain question. Ought the government, or ought it not, to be under the complete control of the possessors of large property? Those of the ministers who think that it ought, with nearly the whole of the Whig aristocracy, will combine with the Tories in a determined resistance to all further extension of popular influences. Those who think that it ought not, together with two-thirds of those members of the House of Commons who now support the ministry, will form a powerful Opposition party, resting upon the people. The contest will then be short and sharp, between the two principles which divide the world, the aristocratic principle and the democratic; and in such a "stand-up fight," he is an indifferent prophet who cannot foresee that the victory

will be with the side where the strength is growing, not with that where it is waning.[14]

The following eighteen months in no way dampened Mill's zeal for the ballot. The increasing debility of Melbourne's ministry merely heightened Mill's expectation of imminent triumph. Difficulties for the Whig government meant opportunities for the radical cause. To Tocqueville he wrote, in January 1837:

> If any ministry would now bring forward the ballot, they would excite greater enthusiasm than even that which was excited for the Reform Bill. But as matters stand, the Whigs' majority is slipping away from them, & nothing will keep the Tories long out of power except either the adoption by the Whigs of a more radical policy, or the rise among the radicals themselves of able & energetic leaders, acting quite independently of the Whigs . . . You will soon see the ballot a cabinet measure, & then reform will have finally triumphed: the aristocratical principle will be completely annihilated, & we shall enter into a new era of government. The approaching session will be next to that of 1830/31, the most important since 1688—& parties will stand quite differently at the commencement & at the close of it.[15]

Bad as the plight of the Whigs might be, Mill was mistaken in his sanguine estimate of what their predicament would bring. The Whigs were indeed rapidly losing ground to the Tories, and the 1837 election, with its attendant corruption and intimidation, left the Melbourne ministry with a slender and uncertain majority. Not that the radicals had reason to rejoice—their contingent in Parliament, none too effective before this date, had also been reduced. Instead of concluding that the government's problems could be solved by taking on board radical policies, Melbourne argued that Whig losses had resulted from concessions already made to the radicals.[16] When Parliament met in late November, Lord John Russell, whose conversion to the ballot seemed a necessary precondition for carrying the measure, emphatically stated his unequivocal opposition to secret voting and any other proposal he considered incompatible with the 1832 Reform settlement. Radical infiltration of the Whig ministry, either in the form of measures or men, failed to materialize.

Mill refused to become despondent. In response to Lord John's "finality" speech, he strongly urged that the ministry had forfeited any claim to radical support. The minimum requirement for continued radical cooperation, according to Mill, had been a government commitment to the ballot.

> We asked them for nothing but to serve themselves. We asked no more in return for their being supported in office, than that they would consent to be kept in it. We asked only that they would propose the Ballot, in the last Parliament in which they can remain Ministers without the Ballot . . . [T]hey [the radicals] have supported Ministers till the time when the Ballot became a vital question to their remaining Ministers on the principles they have hitherto professed. This exact time Ministers have chosen for declaring a degree of enmity to the Ballot, which they have never before expressed: and here, therefore, it is necessary that our support should terminate.[18]

Far from abandoning hope, Mill saw Whig intransigence as a fillip to the radical cause. "If things continue as they are, we shall behold in another session, if not Sir Robert Peel and Lord John Russell, Sir Robert Peel's and Lord John Russell's followers, seated on the same benches, and enthusiastically supporting the same Ministry; while the opposition benches will be occupied by the Radical party."[19] If, on the other hand, the Whigs, following their inevitable defeat in the House (for they could not go on without radical votes), did not join the Tories, the result would be their moving a no-confidence motion against a Tory ministry. The radicals would support such a motion and the formation of a Whig-Radical government would follow.[20] "Such a Ministry would either itself be, or would prepare the way for, that of which the time will soon come, a Ministry of moderate Radicals."[21]

None of the series of events postulated by Mill came to pass. At the close of the decade a shaky Whig administration remained in place. A medley of radicals, dispirited and leaderless, could take comfort only in the fact that the ascent of the Tories had not yet resulted in their supplanting the Whigs. Although George Grote's ballot motion attracted 200 votes in 1838, and 216 in 1839, it was patently clear that the vast majority of those supporting Grote's motion, radicals included, were not ready to make the issue a question of confidence in the Whig government. Before the close of 1839 even Mill grasped that his hopes had been ill-founded, a realization helped along by Lord Durham, the man Mill had cast in the role of savior of the radical cause. Durham, a vain and erratic political figure whose inept performance in Canada in 1838 had won Mill's calculated applause, declined the proffered honor.

Mill's last major political article of the decade, "Reorganization of the Reform Party," appeared in April 1839.[22] Composed at a time when Mill had not yet despaired of Durham's taking command of the radical forces, this essay once again gives voice to his convictions regarding the nature of the contemporary political struggle. He divides political society into two groups, Conservatives and Liberals (radicals and moderate radicals make up the latter). "If we would find . . . the line of distinction between the two parties . . . we must find out who are the Privileged Classes, and who are the Disqualified. The former are the natural Conservatives of the country; the latter are the natural Radicals."[23] Mill deduces that the majority of the middle classes and all of the laboring classes are "natural Radicals."[24] In his view, a shared fundamental objective unites the disparate social, religious, and national groups—merchants, manufacturers, skilled and unskilled laborers, liberal Churchmen and Dissenters, Irish and Scots—that he designates "natural Radicals." Mill sets forth his idea of radicalism's essence.

> One Radical differs from another as to the *amount* of change which he deems necessary for setting what is wrong right: but as to the *kind* of change there is no disagreement: it must be by diminishing the power of those who are unjustly favoured, and giving more to those who are unjustly depressed: it must be by adding weight in the scale to the two elements of Numbers and Intelligence, and taking it from that of Privilege.[25]

From his argument it is clear that Mill's perspective on the character of the struggle had not changed. Neither had his purpose. Yet his ardor on behalf of the ballot as a servant to that purpose had begun to cool. The ballot, while still thought necessary by Mill, no longer occupied such a prominent place in his assessment of what existing circumstances required. "It is not for the Ballot, nor even for the Ballot accompanied by Household suffrage, that the whole force of the Movement party will ever again take the field."[26]

What had brought Mill to this conclusion? In "Reorganization of the Reform Party" he admits that universal suffrage was not, as matters stood, a practicable proposal. The middle classes, particularly with the emergence of the Chartist threat, strongly objected to universal suffrage, for which the working classes had not yet been adequately prepared.[27] Mill therefore judged that the radical politician should take as his motto

"Government *by means* of the middle for the working classes." Until universal suffrage became feasible, the object should be "to govern the country as it would be necessary to govern it, if there were Universal Suffrage and the people were well educated and intelligent."[28] Mill wanted the middle-class electorate to have the protection of the ballot; he also believed that radicalism would have no chance of success unless a large measure of political cooperation between the middle and laboring classes could be achieved. That such cooperation did not then prevail had been made all too clear by the rise of Chartism. Mill was sensitive enough to recognize that Chartist advocacy of the ballot was tied inseparably to the demand for universal suffrage. The Chartists did not care to see an electorate dominated by the middle classes given the protection of secret voting.[29] Mill acknowledged this sentiment.

> It is the opinion of the Operatives, that unless the Suffrage comes down to their own level, anything which enables it to be exercised more independently does them harm. The men of thews and sinews will never give their confidence to a party recommended only by willingness to take from the aristocracy and give to the shopocracy.[30]

The ballot, Mill now saw, was not an issue likely to foster cooperation between working and middle classes.

The tone of Mill's treatment of the ballot had thus shifted considerably. His advocacy of the measure, previously so ardent in character, had become far more subdued. One finds confirmation of this change in a letter he wrote in October 1839. He states that "The ballot though in my opinion necessary, & but little objectionable, is passing from a radical doctrine into a Whig one as will be seen the moment it is carried. It is essentially a juste milieu, middle class doctrine."[31] This assessment of the ballot reflected Mill's altered understanding of the political landscape. There had never been, on Mill's part, a doctrinaire commitment to the principle of secret voting. In the mid-1830s he had pressed the ballot as the single issue best calculated to undermine aristocratic domination. By 1839 developments both inside and outside Parliament had persuaded him that the ballot had lost much of its utility relative to the radical aspirations governing his political activity during the latter half of the decade.

Mill's role of propagandist for the radical cause and his conception of the contemporary political context had shaped his commitment to the ballot in the 1830s. His own standing within the culture of nineteenth-century England was noticeably different by the mid-1850s; by then the political scene too had changed in certain ways. Which category of change—the personal or the political—mattered more in connection with his apostasy on the ballot question?

The precise moment when Mill turned against the ballot cannot be known. John M. Robson has convincingly argued, on the basis of revisions made for the third edition of *A System of Logic* (1851), that Mill's conversion to open voting took place between 1846 (when the second edition of the *Logic* was published) and 1851.[32] There is reason to think it did not occur before the autumn of 1848. In the summer of that year Mill wrote several leading articles on parliamentary reform for the *Daily News*. These articles were prompted, in part, by Joseph Hume's bringing his "Little Charter" before the House of Commons. Hume's package included household suffrage, the ballot, and triennial parliaments. Mill strongly backed this initiative, and said nothing that implied antipathy to the ballot.[33] The first explicit indication we have that Mill no longer considered secret voting necessary or desirable comes in a letter of March 1853, where he announces that the "ballot would be a step backward instead of forward."[34]

The question of Harriet Taylor's influence on Mill during the late 1840s and early 1850s naturally arises in connection with his change of mind on the issue. In the *Autobiography* Mill states that the "hostility to the Ballot" expressed in *Thoughts on Parliamentary Reform* mirrored "a change of opinion in both of us, in which she rather preceded me."[35] Readers might infer from this remark that Harriet had played a prominent part in turning him against the ballot; they might, just as sensibly, surmise that Mill had said as much as he could for her by mentioning that she happened to get there first. In view of the fulsome recognition he heaps on Harriet for showing him the light with regard to other matters, and the fact that Mill is usually not prepared to trust his readers to infer Harriet's general superiority, much should be made of the absence of any declaration, either in the *Autobiography* or in extant letters, to the effect that his wife's arguments were responsible for his rejection of the ballot. The probability is that Harriet reinforced a judgment that Mill had independently arrived at.

All the same, Harriet's strong conviction of the issue's importance almost certainly had a lot to do with the fairly extensive treatment given the ballot in *Thoughts on Parliamentary Reform*. In 1854 she urged her husband to write an article devoted entirely to the subject. Mill demurred.

> I do not feel in the way you do the desirableness of writing an article for the Ed[inburgh] on it. There will be plenty of people to

say all that is to be said against the ballot—all it wants from us is the authority of an ancient radical & that it will have by what is already written & fit to be published as it is [*Thoughts on Parliamentary Reform*]—but I now feel so strongly the necessity of giving the little time we are sure of [Mill believed both he and Harriet were dying of consumption at this time] to writing things which nobody could write but ourselves, that I do not like turning aside to anything else.[36]

Harriet's keen interest in the question no doubt helped persuade Mill that his essay on parliamentary reform should incorporate a substantial critique of secret voting.

Factors outside the home influenced the timing of Mill's rejoining the debate on the ballot after a lapse of nearly fifteen years. The fact that he did not touch upon the topic in the 1840s is hardly surprising, the issue having faded into obscurity in this decade. By the early 1850s, however, the ballot appeared to be making a comeback. Between 1839 and 1846 middle-class reformers had concentrated their attention on the repeal of the Corn Laws. By 1852 the triumph of free trade at last seemed secure and a number of radicals were looking for a new issue to use against the aristocracy. In the early 1850s Richard Cobden and John Bright enthusiastically took up the ballot.[37] Henry Berkeley's ballot motion attracted the support of 172 MPs in 1853. These developments, attesting to the fact that the ballot was again a matter of contention, contributed to Mill's renewed interest in the subject.

Mill's reconsideration of the ballot was markedly affected by the impact of the intervening decade and a half on his own fortunes. In the late 1830s he had been the leading journalistic advocate for a particular brand of radicalism. Long before the mid-1850s the days of Philosophic Radicalism as a distinct political movement had passed. The publication of *A System of Logic* in 1843 and the *Principles of Political Economy* in 1848 had established Mill's reputation as a thinker of the first rank. The authority he now enjoyed was that of a philosopher with public aims in view; in certain respects, this authority lay beyond the reach of the journalist or politician. Whatever he might say on various issues would acquire force simply by virtue of their having been uttered by one of the leading thinkers of his day. His pronouncements on the ballot in the 1850s were those of a political philosopher and public moralist, not those of a radical propagandist.

Mill's sense of his own intellectual and political independence also may have influenced his handling of the ballot in the 1850s. He and

Harriet conceived of themselves as leading the way.[38] They were disciples of no one (indeed, many serious young university men were reverently cutting their intellectual teeth on Mill's *Logic* and *Political Economy*). Mill owed no debts to any party or group, and the authority he possessed stemmed from his remarkable achievements as a philosopher and political economist. Opposition to the ballot set Mill apart from most contemporary radicals, for whom he felt some contempt. In a letter to his wife he notes that the "ballot has sunk to far inferior men, the Brights &c. When it was in my father's hands or even Grote's such trash was not spoken."[39] It is appropriate to cite, in this context, a passage dealing with the ballot that figures in another letter Mill wrote to Harriet. Referring in 1854 to the section on open and secret voting in *Thoughts on Parliamentary Reform*, Mill says that it "might serve to float the volume as the opinion on the ballot would be liked by the powerful classes, and being from a radical would be sure to be quoted by their writers, while they would detest most of the other opinions."[40] The point here is not to argue that Mill's opposition to the ballot sprang from a desire to distinguish his own views from those of commonplace radical opinion; it is to suggest that a defense of open voting, in drawing a line between such opinion and the views of the Mills, carried an ancillary advantage.

To that defense Mill devotes nearly one-third of *Thoughts on Parliamentary Reform*. Readers of the pamphlet quickly discover that Mill has no intention of conceding that his support of the ballot in the 1830s had been mistaken. He states that "secret suffrage, a very right and justifiable demand when originally made, would at present, and still more in time to come, produce far greater evil than good."[41] Mill then briefly refers to the purpose of secret voting and, in keeping with his father's theoretical analysis in the *History of British India*, outlines the circumstances that warrant its adoption.[42] In his father's time the ballot was desirable because the probability of an elector casting his vote in a way consistent with the public interest would have been decidedly enhanced by the protection against intimidation it provided. Such is no longer the case, Mill asserts.

Mill adduces the great change that had occurred in the political and social organization of English society as the principal reason for his rejection of the ballot. In the 1830s, he contends, the upper classes were in complete control of the government. The enactment of the ballot at that time would have been a major gain for reform as "it would have broken the yoke of the then ruling power in the country—the power which had created and which maintained all that was bad in the institutions and the administration of the State—the power of landlords and boroughmongers."[43] He goes on to say that the distribution of political power had substantively changed since the 1830s. The days when the higher classes exercised an exclusive dominion over the political system had passed. The middle classes were no longer subservient to the upper, and the working classes had escaped dependence through an increasing prosperity and improved collective organization. Whereas the ballot should have been accepted in the 1830s as a necessary evil, the progress made by the nation since then not only obviated the need for secret voting but meant that its introduction would now be a retrograde step.

> Thirty years ago it was still true that in the election of members of Parliament, the main evil to be guarded against was that which the ballot would exclude—coercion by landlords, employers, and customers. At present, I conceive, a much greater source of evil is the selfishness, or the selfish partialities, of the voter himself. A "base and mischievous vote" is now, I am convinced, much oftener given from the voter's personal interest, or class interest, or some mean feeling in his own mind, than from any fear of consequences at the hands of others: and to these evil influences the ballot would enable him to yield himself up, free from all sense of shame or responsibility.[44]

From Mill's standpoint, the contribution publicity made to the formation of a sound moral and political character should now take precedence over the desire to protect the elector against coercive influence. Like *Thoughts on Parliamentary Reform*, Mill's personal correspondence conveys his low opinion of the average elector's capacity to recognize the public interest and act accordingly. The introduction of the ballot, he feared, would lead voters to infer that they possessed the suffrage as a matter of right; that they could do with it what they liked; that they need not consider the welfare of their fellow citizens or the community at large when casting their vote. The secret vote, he told George Cornewall Lewis in March 1859, would be falsely interpreted as "a recognition by the State that electors may vote as they please, and are not accountable for their vote as a moral act."[45] The result would be a general lowering of the nation's political morality, together with an increase in bribery, corruption, and the incidence of electors whose conduct at the polls was determined by selfish motives. Mill wishes to insist, above all else, that the act of voting was a public process, public duty, and public trust. The elector's responsibility for the health of the body politic had to be acknowledged and accepted. "There will never be honest or self-restraining government unless each individual participant feels himself a trustee for all his fellow citizens and for posterity."[46] Public voting, Mill held, was one important means of educating the people to understand the significance of the franchise. Of publicity he wrote: "Nothing less than the most positive and powerful reasons of expediency would justify putting in abeyance a principle so important in forming the moral character either of an individual or of a people, as the obligation on every one to be ready to avow and justify whatever he does, affecting the interests of others."[47]

Mill emphasized that public voting had a morally beneficial effect on the individual elector. Publicity ensured that those who acted contrary to the prevailing opinion would do so on the basis of moral and intellectual conviction. "To be under the eyes of others—to have to defend oneself to others—is never more important than to those who act in opposition to the opinion of others, for it obliges them to have sure ground of their own."[48] This notion, of course, is consonant with a central theme of *On Liberty*, published the same year as *Thoughts on Parliamentary Reform*. He wanted the nation's political institutions, wherever possible, to cultivate in its citizens an active, energetic, and independent character. In defending public voting he saw himself as serving this cause.

> The moral sentiment of mankind, in all periods of tolerably enlightened morality, has condemned concealment, unless when required by some overpowering motive; and if it be one of the paramount objects of national education to foster courage and public spirit, it is high time now that people should be taught the duty of asserting and acting openly on their opinions.[49]

Voting at parliamentary elections, in Mill's view, did not fall into the category of self-regarding actions. The interests of others were at stake when individual electors expressed their political judgment at the polls. From this it followed that open voting should be the norm irrespective of whether the suffrage was limited or universal in extent. In the former case, electors acted as trustees for nonelectors; just as the responsibility of MPs to the public at large required that voting in the House of Commons be open, so voters making up a circumscribed electorate had to exercise their responsibility publicly.[50] In the latter case, that of universal suffrage, the concept of the vote as trust remained in force. The interests of each elector would be affected by the interests and agency of all others, and the well-being of the entire political community would still depend on each voter discharging his duty as a public trustee. Publicity encouraged fulfillment of this obligation. "The universal observation of mankind has been very fallacious, if the mere fact of being one of the community, and not being in a position of pronounced contrariety of interest to the public at large, is enough to ensure the performance of a public duty, without either the stimulus or restraint derived from the opinion of our fellow-creatures."[51]

Before assessing the significance and implications of Mill's argument, something should be said about the impact of his intervention in the wider debate over the ballot. As far as opponents of the measure were concerned, the content of Mill's argument mattered less than the fact that a prominent and influential radical, in time past an ardent supporter of secret voting, had now turned against it. Such a development plainly gave additional weight to their defense of open voting. Mill's polemic, however, did not fundamentally affect the substance of that defense, which had evolved over a period of several decades.[52] In elaborating a constitutional justification of public voting, the opponents of the ballot had always given central place to the trust concept of the vote. Although Mill's articulation of this regulative principle bolstered its authority, the crux of the trust argument remained unchanged.

The major gain for the antiballot forces was indirect. The publication of *Thoughts on Parliamentary Reform* damaged the pro-ballot movement. Many advanced liberals looked to Mill for intellectual and political guidance during the mid-Victorian decades. Before 1859 most men of this sort were disposed to support the ballot as a reform consistent with their political objectives. Mill's repudiation of secret voting might well raise doubts in their minds regarding its desirability. In giving evidence before the Select Committee on Parliamentary and Municipal Elections in 1869, William Latham, a Cheshire magistrate and active Liberal who had expressed his opinion in favor of the ballot, was asked whether he had always been a supporter of secret voting. He replied in the negative, saying that "I was a disciple of Mr. John Stuart Mill, and I believed that the ballot was an evil."[53] Advocates of the ballot acknowledged that Mill's condemnation of secret voting had inflicted serious harm on their cause. John Bright admitted that "Mr. Mill's opposition is very unfortunate."[54] Henry Berkeley, the chief sponsor of the ballot in the House of Commons, asserted that "Mr. Mill and Mr. [Robert] Lowe [whom Berkeley believed responsible for the antiballot leading articles in *The Times*] are the two worst opponents of the Ballot."[55] Many pro-ballot pamphlets of the 1860s, such as Henry Romilly's *Public Responsibility and*

Vote by Ballot (1865), G.J. Holyoake's *A New Defence of the Ballot* (1868), and the anonymously authored *Mr. John Stuart Mill and the Ballot* (1869), were written with the overt purpose of rebutting Mill's case against secret voting.

The reaction to Mill's recantation casts but a dim light on the genesis of his conversion to open voting or the plausibility of what he said to justify his change of mind. Mill himself implies that the mid-Victorian public would have had no exposure to his plea for publicity in the giving of the vote had he not concluded that electoral conditions had changed dramatically since the 1830s. The validity of this plea therefore cannot be judged simply on the basis of its speculative cogency; it must be assessed relative to an appraisal of conditions on the ground and the degree to which they were different in the 1850s from what they had been two decades earlier. Few students of Victorian elections have closely considered Mill's change of position on the ballot. A prominent exception is D.C. Moore, who maintains that Mill's rejection of secret voting can indeed be understood as arising from his perception of an altered electoral environment.[56] The key concept in Moore's examination of the electoral arena is the "deference community." The hierarchical structure of this community, according to Moore, determined the nature of electoral behavior in the localities (his study is largely confined to the counties) during the middle years of the nineteenth century. Tenants, he says, generally voted with their landlord because the latter was seen as the legitimate source of political authority in the community, an authority rooted in his social and economic position in the countryside. Moore argues that both James and John Stuart Mill acknowledged the rightful claims of a hierarchical "deference community." Observing that James Mill frequently touted the merits of legitimate influence, Moore urges that the elder Mill's advocacy of the ballot originated in his commitment to the creation of a genuine "deference community" uncorrupted by the exercise of illegitimate influence. The natural leaders of society—the most talented and benevolent of the middle ranks—were not deferred to by the electorate because coercive and corrupt practices got in the way.[57]

The younger Mill's ultimate opposition to the ballot, Moore suggests, resulted from his perception that "the discipline exercised by the traditional agencies of social control was becoming weaker. But, he [Mill] contended, these agencies could be preserved if open voting were retained."[58] The implication is that Mill, without grasping fully the nature of the disciplinary agencies involved, had come to approve of the effects they produced. The deference networks responsible for this discipline were beginning to decay, however, owing to advancing urbanization and the resolve of new elites to challenge the old. Moore does not pretend that Mill understood this process, but he intimates that Mill's worry about the number of voters being influenced by selfish personal and class interests arose in response to the decline of deferential politics. This worry, in turn, underlay his condemnation of the ballot.[59]

Moore's commentary on J.S. Mill and the ballot misconstrues the position Mill set forth in *Thoughts on Parliamentary Reform*. To be sure, Mill does assign a disciplinary function to "public opinion." For Mill, however, "the traditional agencies of social control" had nothing to do with the legitimate influence of "public opinion." Those agencies, in his view, embodied the power of landlords and employers to bully their dependents, and he applauded the weakening of this power. What concerned him was the need to create effective new moral agencies to counteract the tendency, as he saw it, of electors to vote on the basis of selfish considerations now that they were no longer subject to dictation.

How well-informed was Mill about the forces shaping electoral behavior in early and mid-Victorian Britain? In the 1830s he had insisted that intimidation and corruption were rife and that secret voting was the only remedy; when he took up the ballot question in the 1850s, he adhered to this view of the 1830s but maintained that electoral changes had eliminated the need for the ballot. If the problems so evident to Mill in the 1830s had been solved in less than a generation, without recourse to secret voting, why should he nonetheless have thought in the 1850s that the ballot ought to have been adopted two decades earlier? Neither in the 1830s nor in the 1850s was Mill's position on the issue decisively swayed by an astute appreciation of electoral conditions. In the 1830s Mill had held that intimidation was rife in the English counties. However flawed may be his interpretation of Mill's politics, D.C. Moore has demonstrated a commanding knowledge and understanding of county politics and elections in the thirty years between 1830 and 1860. His *Politics of Deference*, a monumental study, effectively makes the case that coercion did not play a significant role in deciding the votes of the English tenant farmers who made up the bulk of the county electorates. This finding applies to the 1830s no less than to the decades that followed. As for Mill's comments about how much had changed by the 1850s, they do not convince. The eviction of Welsh and Irish tenant farmers for defying the electoral instructions of their landlords was not uncommon in the decades between the First and Second Reform Acts.[60] In some boroughs publicans and shopkeepers with the vote continued to feel pressure from nonelectors who happened to be their customers.[61]

Advocates of the ballot saw it as an antidote for bribery as well as intimidation, as it would deny those in the business of purchasing votes the means of knowing whether they had got value for their money. Venality at elections did not diminish between the late 1830s and the mid-1850s. The general election held shortly before Mill wrote *Thoughts on Parliamentary Reform* was among the most venal of the age. T.J. Nossiter, author of *Influence, Opinion and Political Idioms in Reformed England*, goes so far as to say that "it was this election, which, arguably, marked the high water mark of corruption in reformed England."[62] The extent of bribery and intimidation at the 1852 general election colored Cobden's decision to launch a new ballot campaign.[63] Shortly after the election Sir James Graham, an opponent of secret voting, wrote Lord Dunfermline to tell of his fear that the extensive corruption and coercion evident in many constituencies would produce an intensification of the public demand for secret voting.[64] In 1853 Lord Palmerston (also antiballot) argued in a letter to Lord John Russell that "the reform which most men call for is not so much a general revision of our representative organization, as a remedy for those abuses of bribery & corruption which were exposed by the Proceedings of the Election Committees last session."[65] In 1854 the Aberdeen Coalition, concluding that a legislative response to the revelations of the 1852 general election was called for, enacted a Corrupt Practices Act (which proved to be ineffectual). Of course the mode of voting alone was not responsible for the practices disfiguring mid-nineteenth-century British elections. The continued prevalence of such practices, however, should have caused Mill to hesitate before mounting a vigorous defense of open voting.

Electoral conditions had changed far less between the 1830s and mid-1850s than had Mill himself. The nature of the change in Mill is best summed up in a passage from the *Autobiography*.

> In England, I had seen and continued to see many of the opinions of my youth obtain general recognition, and many of the reforms in institutions, for which I had through life contended, either effected or in course of being so. But these changes had been attended with much less benefit to human well being than I should formerly have anticipated, because they had produced very little improvement in that which all real amelioration in the lot of mankind depends on, their intellectual and moral state . . . I am now convinced, that no great improvements in the lot of mankind are possible, until a great change takes place in the fundamental constitution of their modes of thought.[66]

Mill's commitment to the ballot in the 1830s was of a totally different order from his commitment to open voting in the 1850s and thereafter. In the 1830s Mill the political sectarian looked to the ballot as an indispensable radical tool for lancing the boil of aristocratic predominance. By the end of the decade the measure had lost much of its significance for him because the occasion for its usefulness within this context of political struggle had passed. Over the next fifteen years Mill's bent toward political activism remained largely dormant while he climbed the speculative heights and attained an unrivalled intellectual eminence. Small wonder he came to hold the view expressed in the *Autobiography*— moral and intellectual progress would not follow upon much needed political change but was rather a precondition of that change. Mill's unrequited commitment to the ballot as an instrument of political power had been superseded by a misguided commitment to public voting as an instrument of political virtue.

CHAPTER SIX

Mill and the Problem of Party

More potent at certain times than others, party yet held a significant place in British political life throughout the Victorian period. Few students of nineteenth-century British politics, however, would look to J.S. Mill, the most influential political thinker of his time, for insights into the role of party. Party, it is generally thought, had no standing in Mill's conception of a healthy political order. In *The Elements of Politics*, Henry Sidgwick says that "Mill . . . hardly seems to contemplate a dual organisation of parties as a normal feature of representative institutions."[1] A.H. Birch, in his *Representative and Responsible Government*, asserts that Mill "simply ignored the existence of political parties."[2] Dennis Thompson's study of the structure of Mill's mature political thought devotes some three pages to Mill's attitude toward party government, the author concluding that he was hostile to party and considered it unnecessary "for effective, stable democracy."[3] A recent assessment of Mill's political theory, Nadia Urbinati's *Mill on Democracy*, echoes Thompson's judgment. Mill, Urbinati states, condemned party "for restricting competition among individuals and engendering mediocre electoral choices."[4] These estimates rest on an unimpeachable textual foundation. Mill's major political treatise, *Considerations on Representative Government*, says remarkably little about parties, and this little imputes no constructive influence to them. One of the many great virtues of Thomas Hare's plan of personal or proportional representation, Mill argues, was its capacity for ensuring the representation not of "two great parties alone" but of every significant "minority in the whole nation."[5] Personal representation would enable many talented men of "independent thought," men who had "sworn allegiance to no political party," to win election to Parliament.[6] Yet to concentrate on the meager treatment of party in *Representative Government* and on the negative character of this treatment yields an incomplete understanding of Mill and the question of party.

The author of *Representative Government*, a work written and published during the Palmerstonian ascendancy, found a void where he looked for parties that stood for a coherent set of political precepts. In the Preface to *Representative Government* Mill declared: "both Conservatives and Liberals (if I may continue to call them what they still call themselves) have lost confidence in the political creeds which they nominally profess, while neither side appears to have made any progress in providing itself with a better."[7] Mill's criticism is directed at the existing party system, not at the principle of party. *Representative Government* posits an institutional framework conducive to establishing and maintaining a progressive democratic polity. Mill did not conceive of party in institutional terms, and its absence from *Representative Government* can be explained in part by his notion of what he thought party should be. He did not envisage a political system free from party conflict; rather, he wanted the petty party warfare of his own day to give way to a fruitful political antagonism between systems of belief. John M. Robson has demonstrated the important place antagonism occupies in Mill's political thought.[8] Organized antagonism implies party, in some form, and Mill's conception of the House of Commons as "not only the most powerful branch of the Legislature," but "also the great council of the nation; the place where the opinions which divide the public on great subjects of national interest, meet in a common arena, do battle, and are victorious or vanquished,"[9] indicates the need for parties of a certain type.

The centrality of party to Mill's political activity during the 1830s has been lucidly examined by Joseph Hamburger.[10] Mill's political articles of this decade tried to build support for the formation of a genuine radical party dedicated to the extinction of aristocratic government and the democratization of British institutions and society. Hamburger has ably analyzed the ideological foundations of this activity. Mill, Hamburger demonstrates, saw many Whigs as little more than Tories in disguise. The political warfare engaged in by Whigs and Tories was largely intended to obfuscate the true line of battle, the aristocracy being on one side of this line and the "people" on the other. Whigs and Tories alike were wedded to the preservation of aristocratic domination, and averse to parting with any of the power they possessed. Mill foresaw most Whigs joining up with the Tories as the political environment became increasingly radicalized. About both, the radicalization and the Whig-Tory merger, he was mistaken. The failure of Mill and like-minded

Philosophic Radicals to create an effective alternative to the Tories and the Whigs, his eventual realization that this aspiration could not be reconciled with the prevailing realities of the political order, and the absence of any comparable activity on his part in subsequent years have been taken to mean that his commitment to party in the thirties expired with his growing disillusionment. By 1840 Mill had indeed grasped the futility of propagandizing in the service of a radical cause that was disintegrating and losing what support it had gained. The realignment of the party system he had worked to promote was not going to happen. It does not necessarily follow that he then gave up the ideal of party to which he had adhered in the 1830s.

For Mill party was the means of giving organized political expression to ideological commitment. In his "Notes on the Newspapers" (1834), he observed that "No body of men ever accomplished any thing considerable in public life without organized co-operation."[11] The nature of the party is defined by what it seeks to achieve; its legitimacy, in Mill's eyes, rests on the distinctive interests, principles, and convictions it embodies. He does not see party as an institution giving rise to loyalties and conflicts that exist independently of substantive issues and interests. He grants the presence of such loyalties and conflicts relative to existing parties, but not their legitimacy. Taking issue with the conception of party set forth by Francis Jeffrey in the *Edinburgh Review*, which appeared to sanction violation of principle when required to further the ambitions of the party for office, Mill (in a *Westminster Review* article of 1824) declared:

> No one is more sensible of the necessity of concert than ourselves. Not that sort of concert which consists in speaking and voting on one side, thinking and feeling on the other—but a concert which involves no sacrifice of principle—a concert for mutual aid among those who agree, without imposing fetters upon those who differ; a concert in short, not for men, but for measures.[12]

In the 1830s and thereafter, Mill had no doubt that parties of this sort were a good thing. Moreover, in 1839 he went so far as to say that there should be two such parties, and no more. "There may be many *coteries* in a country, but there can be only two parties."[13] These two, of course, were the Conservative and the Liberal parties.

Mill conceived of the Conservative party as generally made up of those who already had power, wealth, and influence—those who had a vested interest in preserving the nation's established institutions. But a

Conservative party truly worthy of its name must be more than the instrument of powerful material and institutional interests. The commitment of Conservatives to permanence, order, and stability should rest on a firm intellectual foundation. They should understand what these ideas signified and promote specific measures consistent with their broader commitments; they must "systematize and rationalize their own actual creed."[14] In Mill's view, the problem with the early and mid-Victorian Tory party was that most of its members had no intellectual grip on the principles that should have governed its conduct. The bulk of the Tories were moved not by principles, but by blind passion, prejudice, and instinct.[15] Writing to Thomas Hare in 1860, Mill said that the Conservative party was "not only the least powerful, but the silliest party. It has been left behind by all its able men, and the others are daily shewing that of all politicians the Conservatives are the least alive to any real *principles* of conservation."[16] Mill concluded that the Conservative party was constitutionally incapable of fulfilling the purpose such a party was meant to serve. "The Conservatives . . . by the law of their existence [are] the stupidest party . . . and it is a melancholy truth, that if any measure were proposed, on any subject, truly, largely, and far-sightedly conservative, even if Liberals were willing to vote for it, the great bulk of the Conservative party would rush blindly in and prevent it from being carried."[17]

What would a Liberal party worthy of its name look like? Mill's Liberal party was to be the party of reform, progress, and movement. Its function—in the broadest sense—was to attack political, religious, and economic privilege, and to foster a political and social environment favorable to the full development of the individual's moral and intellectual capabilities. In an 1865 election address Mill set out to differentiate the Tory from the Liberal.

> A Tory is of opinion that the real model of government lies somewhere behind us in the region of the past, from which we are departing further and further. Toryism means the subjection and dependence of the great mass of the community in temporal matters upon the hereditary possessors of wealth, and in spiritual matters to the Church, and therefore it is opposed to the last moment to everything which could lead us further away from this model . . . The Liberal is something very different from this. The probability is, that we have not yet arrived at the perfect model of government—that it lies before us and not behind us—that we are too far from it to be able to see it distinctly except in outline, but

that we can see very clearly in what direction it lies—not in the direction of some new form of dependence, but in the emancipation of the dependent classes—more freedom, more equality, and more responsibility of each person for himself.[18]

If Mill considered the Tory party of Derby and Disraeli far inferior to what a Conservative party ought to be, he also found the Liberal party of Russell and Palmerston seriously wanting.

The state of national politics in the quarter century after 1840 seldom aroused in Mill enthusiasm sufficient for explicit declarations linking principle and party. An exception occurred in the summer of 1848, when Joseph Hume brought before the House of Commons his "Little Charter," a program of reform calling for household suffrage, the ballot, triennial parliaments, and a greater equality of electoral districts. The issue of parliamentary reform appeared to be reviving, and Mill responded with three leaders on the subject in the *Daily News*.[19] In these articles he condemned the arguments against reform offered by Lord John Russell, and urged a pressing of the question as a means of setting apart those seeking democratic change from those who want "no change at all, or . . . such changes only as would make no difference in the spirit of the government." Above all, he wanted to encourage sincere reformers to make clear the principles upon which they acted.

> One lesson the consistent supporters of reform may take to themselves—a lesson which becomes more important in proportion as the contest ceases to be a mere mock fight and becomes a serious conflict of opposing reasons. Their practical conduct as politicians necessarily partakes of compromise. Their demands and systematic aims must often fall short of their principles. But let them not therefore cut down their principles to the measure of their demands. If they do, they lose far more in vigour of argument, and in the imposing influence of a sense of consistency and power, than they can possibly gain in charming away the fears of those who would, but dare not, follow them. Let them disclaim nothing which is a legitimate consequence of their principles. Let them tell the truth—when it is the truth—that their private opinion goes further than their public demands, and that if they ask less than what their principles would justify, it is not because they fear to avow, or are unable to defend, their principles, but because they think they are doing more good by uniting their efforts with those of others to attain a nearer object, and one more immediately practicable.[20]

Mill here advocates a party contest based on "a serious conflict of opposing reasons." Genuine reformers had a responsibility to focus on issues calculated to advance this conflict, and to express forcefully and consistently the principles that guided their conduct. Mill's position on this matter remained much the same as it had been a decade earlier. Two decades after the *Daily News* articles of 1848 this stance was again in evidence.

The political atmosphere of the Palmerstonian ascendancy discouraged Mill from speaking to the problem of party. The Liberals were doubtless to be preferred to the Tories, but both appeared to lack the capacity to hammer out the distinctive set of directive principles necessary to give purpose, system, coherence, and unity to party action on a range of issues. "Without presuming to require from political parties such an amount of virtue and discernment as that they should comprehend, and know when to apply, the principles of their opponents, we may yet say that it would be a great improvement if each party understood and acted upon its own."[21] Believing principles to be the essence of party, and finding them largely absent from both existing parties, Mill considered the political order of the Palmerstonian ascendancy to be fundamentally nonparty in nature.[22] His endorsement of Hare's scheme of personal representation can be more fully appreciated when viewed in this context.[23] The election of a significant number of able and articulate men—individuals who cared about principle and were independent of existing parties—would help overcome the mediocrity and sterility of the mid-Victorian parties, and perhaps lay the groundwork for a genuine party system. Not only would conflict at the parliamentary level come to reflect principled differences, but party organizers, compelled to respond to a method of representation that gave each voter an opportunity to support an attractive nonparty candidate, would strive to bring forward candidates of a quality superior to those usually nominated.

> They [local party leaders] could no longer count upon bringing up the whole strength of the party, to return any professed Liberal or Conservative who would make it worth their while. An elector even of their own party, who was dissatisfied with the candidate offered him, would not then be obliged to vote for that candidate or remain unrepresented . . . [H]is leader would be under the necessity of offering him someone whom he would consider creditable, to be secure of his vote. It is probable that competition would spring up among constituencies for the most creditable candidates, and that the stronger party in every locality . . . would be anxious to bring forward the ablest and most distinguished men

on their own side, that they might be sure of uniting the whole of their local strength, and have a chance of being reinforced by stray votes from other parts of the country.[24]

Thus Mill's hearty support for Hare's proposal stemmed in part from a conviction that its adoption would encourage the formation of a party system in which ability and principle would receive apt recognition.

It was not until the latter half of the 1860s that Mill found himself in a position to contribute in a direct way to the creation of such a party system. In the preceding quarter-century he had worked to influence the way men thought about themselves, their society, and their institutions, doubtless hoping that this influence would ultimately have a salutary effect on the political conduct of his fellow-citizens. His return to Parliament in 1865 as one of Westminster's two MPs, however, moved him to the center of the nation's affairs.[25] Mill's standing for Westminster did not spring from a reassessment of the potential of the existing party system. Out of a sense of public duty he accepted an invitation from a group of Westminster electors to come forward in the Liberal interest. He did so on the understanding that he would not spend money or canvass in his own cause. He also insisted upon complete freedom to act as he thought best should he be elected.[26] Having asserted in *Representative Government* that an infusion of men of independence and ability would greatly enhance Parliament's effectiveness as a governing body, Mill could not in good conscience decline such an invitation.

The year 1865 saw both Mill's election to the House of Commons and the death of Palmerston. The former development alone meant Mill would encounter practical issues tinged with party implications; the latter meant that he would do so within a highly mutable political context. Palmerston's death released forces bearing on the party system that his presence on the political scene had kept in check. Gladstone now became the Liberal leader in the House of Commons (an aged Russell, who sat in the Lords, held the premiership). Mill wanted Gladstone to be the future leader of an advanced Liberal party, and his rise revived Mill's interest in the question of party.

Palmerston, who died in October, was still at the helm when the general election of 1865 occurred. At the time of the election Mill voiced his attitude to the Liberal and Conservative parties. Speaking to a meeting of Westminster electors on 8 July 1865, he stated that

> He could not look forward to any time in the history of this country when he should not think any Liberal Ministry preferable to any

Conservative Ministry. Whatever the shortcomings of a Liberal party or Government might be, they did not bear in their very names the profession of wishing to keep things as they were. Their name implied that they wished to improve them; and although between the least liberal of Liberals and the most liberal of Conservatives there might be only a little difference, a short distance, still it should be ever borne in mind, and seriously remembered, that this least liberal of Liberals was surrounded by those who were far better men than himself, politically speaking, while this most liberal of Conservatives was surrounded by men who, politically speaking, were far worse than himself . . . If he were returned to Parliament, what he should do, and that which he should recommend others to do, would be to vote for any Liberal Government on questions between them and the Tory Government, but he should not let himself be muddled under the pretence of keeping a Liberal Government in; therefore he would advise the independent Liberals always to vote as they thought best, and let the Government or Ministry shift for themselves, and take their chances of whatever might be the result of a full and free discussion.[27]

In Mill's view, therefore, the electors of Westminster who supported him were voting for an "independent" Liberal.

In late 1865 and early 1866 Mill commented on the condition of the Liberal party in letters to various correspondents. Replying to John Plummer, a working-class journalist who had described the party as "rapidly approaching a state of complete disorganisation," Mill declared, in late November, that if such was indeed the case "the conclusion I should draw would be that it is time for it to dissolve, and organise itself anew on some better basis."[28] Six weeks after writing to Plummer he broached the subject of parliamentary reform in a letter to Thomas Hare. Mill noted that "Most of the liberal members are not real reformers, and only vote for any reform because they are obliged, and in the hope of getting rid of the question."[29] As a Member of Parliament Mill hoped to take part in a purification of the Liberal party that would excrete from its ranks those whose liberalism was of a strictly nominal character. Writing to Theodor Gomperz in August 1866, Mill acknowledged that his acceptance of a seat in Parliament betokened a "sacrifice of time and energies that might be employed on higher work." He added: "Time will show whether it was worth while to make this sacrifice for the sake of anything I am capable of doing towards forming a really advanced liberal party which, I have long been convinced, cannot be done except in the House of Commons."[30]

Mill discusses his conception of his role in Parliament in the *Autobiography*. "When I had gained the ear of the House . . . the idea I proceeded on was that when anything was likely to be as well done, or sufficiently well done, by other people, there was no necessity for me to meddle with it." Hence he reserved himself "for work which no others were likely to do," and, he adds, most of his "appearances were on points on which the bulk of the Liberal party, even the advanced portion of it, either were of a different opinion from mine, or were comparatively indifferent."[31] While thus emphasizing the independent character of his actions in Parliament, Mill at the same time makes his conduct part of an advanced liberal nexus. "The same idea, that the use of my being in Parliament was to do work which others were not able or not willing to do, made me think it my duty to come to the front in defence of advanced Liberalism on occasions when the obloquy to be encountered was such as most of the advanced Liberals in the House, preferred not to incur."[32]

Mill believed he could best contribute to the formation of an advanced Liberal party by acting as an independent agent. He made no claim to an exclusive possession of bona fide Liberal principles, knowing that such a claim would impede rather than further his purpose. Nor did he seek to rally a party round himself. He understood his unfitness for leadership, and knew that the political arena in which he had to function would be impervious to such an undertaking. Instead he used his presence in the House of Commons to bring forward the principles he thought should be at the core of an advanced Liberal party. He did not explicitly define these principles in party terms because there was no existing party to which they could be attached. His task was to give life to the principles themselves in the hope that they would ultimately constitute the essence of a Liberal party ready to embrace them.

Circumstances seemed to favor this project. With the death of Palmerston and the ensuing uncertainty within Liberal ranks, it appeared that the party's future direction would be markedly influenced by the House of Commons in which Mill had a place. An independent Liberal of eminent reputation and authority could reasonably hope to put his imprint on the reshaping of the party.

The emergence of Gladstone as the dominant figure on the political scene raised Mill's expectations, while also limiting his independence so long as a Liberal government was in office. Gladstone's earnestness and restless intelligence, the evidence he had given of a capacity for growth as a statesman, and his indisputable grip on the attention of the House and the public alike account for Mill's fixing upon him as the natural leader of an advanced Liberal party. Yet the hostility to Gladstone harbored by a number of Whigs and moderate Liberals made his ascendancy insecure, and Mill was keen to keep his distance from actions tending to undermine Gladstone's precarious position. This concern goes far to explain his warm endorsement of the government's timid Irish Land Bill,[33] which Mill himself later described as an "extremely mild measure,"[34] and the strong support he expressed for Gladstone's temperate Reform Bill.[35] The resignation of the Liberal government in June 1866 upon the failure of this Reform Bill, and the coming into office of Derby and Disraeli, removed this inhibition, and gave Mill ample latitude for taking up unequivocally radical positions on the Jamaica question, women's suffrage, and Irish land.[36] In doing so, he gave practical and instructive expression to principles he deemed intrinsic to the cause of advanced liberalism. His object, in party terms, was to promote the formation of an advanced Liberal party. If Mill had his way, Gladstone would lead, but not dominate, this party.

Mill warmly and frequently praised Gladstone during the 1868 campaign, the latter's first as leader of the Liberal party. Addressing a meeting of his Westminster constituents in late July, Mill declared that the "present leader of the popular party sincerely desires to do for the people ... the best that can be done," and noted that whatever Gladstone himself did was "incomparably well done."[37] On another occasion later in the year, he told an election meeting that "the one statesman in this country who, perhaps, more than any other within living memory has the confidence of the people is Mr. Gladstone."[38] In mid-November Mill "called upon the electors to support Mr. Gladstone for the very reason for which he was called unsafe. They wanted a Minister who would do things merely because they were right, and who would not mind risking a few votes for his party, if by that risk he could do right and effect a great object."[39]

Mill thus fought his unsuccessful campaign of 1868 as a partisan supporter of Gladstone.[40] Not only had Gladstone been largely responsible for forcing the question of parliamentary reform upon a grudging Parliament; he had also rallied his party in 1868 by proposing the disestablishment of the Irish Church. In Mill's view, Gladstone was the one politician of national stature able and willing to place principle at the center of political discussion. Mill fully appreciated the importance of effective leadership for a party committed to substantial change, and he looked to Gladstone to provide that leadership.

Mill's loyalty, however, was not to the Liberal party or its leader. His allegiance was to an idea of party, and his active support could be counted upon only so long as Gladstone and those who followed him held the promise of realizing this idea. Although Mill's hopes for the future of the Liberal party rested heavily on Gladstone, his confidence in the Liberal leader had its bounds. Gladstone could not play the part Mill wished to assign him without a sizeable corps of dedicated advanced Liberals at his back.

> What the country has to look for is that . . . [Gladstone's] majority shall be more steadfast to genuine Liberal principles. We do not want men who cast reluctant looks back to the old order of things, nor men whose liberalism consists chiefly in a warm adherence to all the liberal measures already passed, but men whose heart & soul are in the cause of progress, & who are animated by that ardour which in politics as in war kindles the commander to his highest achievements & makes the army at his command worth twice its numbers; men whose zeal will encourage their leader to attempt what their fidelity will give him strength to do. It would be poor statesmanship to gain a seeming victory at the poll by returning a majority numerically large but composed of the same incompatible elements as the last.[41]

Such a "seeming victory" is what Mill took the Liberal electoral triumph of 1868 to be. Consoling Edwin Chadwick following his defeat at Kilmarnock, Mill observed that the "new candidates of advanced opinions have been defeated everywhere. Not one working men's candidate (whether a working man himself or not) and not one of the University Liberals has been returned."[42] He informed an American correspondent of "the defeat of the radical party throughout the country."[43] A passage he decided to delete from a January 1869 letter to W.T. Thornton—Mill's friend and former colleague at India House—reveals much about Mill's provisional assessment of Gladstone and the Liberal party in the aftermath of the election.

> The composition of the Ministry is much what we would have expected from the composition of Parliament. Gladstone has evidently interpreted the elections as indicating that the advanced section of Liberals is not strong in the electoral body & he has therefore given the lion's share to the backward section, bestowing only minor appointments on the radicals . . . Gladstone has perhaps

something of the deference of a *novus homo* for the old nobility & he may very reasonably think that the advanced liberals will be content if anything considerable is done for their opinions, while the others must have office to obtain their consent to *any* measures of a radical complexion. I never felt more uncertainty about the immediate future of politics; but I do not doubt that after a few years, & perhaps even at the next general election, the working classes will feel & use their strength; though probably they will not use it fully until the obstacles have been removed to a junction of the Conservatives of both sides of the House against them.[44]

Mill's correspondence of this period frequently alludes to the relation of the working classes to the Liberal party. His reform speeches of 1866 and 1867 made abundantly clear his conviction that the future health of the political order required a comprehension of working-class interests and opinions within the representative system. On the second reading of the 1866 Liberal bill he asserted:

> Every class knows some things not so well known to other people, and every class has interests more or less special to itself, and for which no protection is so effectual as its own . . . I claim the benefit of these principles for the working classes. They require it more than any other class. The class of lawyers, or the class of merchants is amply represented . . . [because] a successful lawyer or merchant easily gets into Parliament by his wealth or social position . . . but no constituency elects a working man, or a man who looks at questions with working men's eyes. Is there, I wonder, a single member of this House who thoroughly knows the working men's views of trades unions, or of strikes, and could bring these subjects before the House in a manner satisfactory to working men?[45]

The 1867 Reform Act had enfranchised a large proportion of the urban working classes. The Liberal party, Mill held, now had a duty to help working men gain election to the House of Commons. Both Gladstonian leadership and an active rank and file with a visible working-class presence were essential for the transformation of the Liberal party that Mill had in mind.

The 1868 general election showed that Liberal constituency organizations were not ready to bring working-class leaders into the fold. Mill told Plummer that the "Liberal party will have cause to repent of not having adopted the best leaders of the working men and helped them to seats."[46] Believing that the Liberals could not survive as a powerful party without working-class support, Mill argued that the working classes should use their strength to secure "an equal voice with the Liberals of the higher & middle classes in the choice of Liberal candidates. Where a place returns two members, one of these should be a candidate specially acceptable to the working classes; where there is but one, he shd be selected in concert by both sections of Liberals."[47] The existing state of the Liberal party left it a vast distance from where it should be; Mill, in the interest of reducing this distance, was prepared to sanction action hardly compatible with a conventional understanding of party loyalty. Writing in early 1870 to George Odger, the working-class politician, Mill observed: "It is plain that the Whigs intend to monopolise political power as long as they can without coalescing in any degree with the Radicals. The working men are quite right in allowing Tories to get into the House to defeat this exclusive feeling of the Whigs, and may do it without sacrificing any principle."[48]

The standard by which Mill judged a party claiming for itself the watchword of reform in the late 1860s is consistent, in spirit if not in detail, with the test he applied in 1839 in his article "Reorganization of the Reform Party." There Mill had made an impassioned plea that justice be done to the laboring classes.[49] Allowing that the political conditions of the late thirties did not admit the feasibility of universal suffrage, Mill insisted that a liberal statesman leading a party of reform "*must redress the practical grievances of the working classes*," and that the "motto of a Radical politician should be, Government *by means* of the middle for the working classes."[50] Although universal suffrage had not arrived in 1867, a substantial measure of working-class enfranchisement had been granted. Justice to the laboring classes and to the public interest now required not government by "the middle for the working classes," but government by the middle and working classes for the nation.[51] A Liberal party that failed to act on this imperative could not be a true party of reform. As in the late 1830s, so in the late 1860s, Mill saw the extinction of Whig influence as essential to the radicalization of the Liberal party. In the thirties circumstances had not been amenable to the substitution of working-class influence for that of the Whig aristocracy; by the late sixties Mill considered a junction of middle and working classes within a radical Liberal party both possible and necessary.

Mill's attenuated discussion of party in *Representative Government* barely signals the approach to the problem he would bring forth in the late 1860s. To the parties of the age of Palmerston he had grave objections. Better to have no parties at all than to have parties unable to justify their existence on the basis of a distinctive and coherent set of principles. Prepared to acknowledge the existence of the parties of the Palmerstonian ascendancy, he was not ready to concede their legitimacy. Indeed, it is possible to read *Representative Government* as a treatise whose institutional recommendations were designed to limit the influence of such parties. Yet the adoption of these recommendations would not have hindered parties whose members joined together to secure objectives defined by a shared ideological perspective. Dennis Thompson has forcefully argued that the specific reforms proposed by Mill derive from the latter's commitment to the precepts of participation and competence, each of which serves an educative and protective function.[52] Mill's conception of party did not clash with this commitment. Mill's Liberal party, through its efforts to give direct representation to the working classes, would both extend the range of participation and enhance the quality of that participation. The priority assigned to principle would presumably raise the level of political discussion, inject a strong sense of purpose into the decision-making process, and give to the implementation of policy a greater measure of efficiency and competence. Party, as understood by Mill, would provide civic education for the masses and protection for the public interest. In demonstrating the right relation between theory and practice, Mill's Liberal party would help construct a rational foundation for the conduct of English politics. This ideal of party, as Mill knew, would be hard to realize. In the 1830s, and again in the late 1860s, he nonetheless saw opportunities to reduce appreciably the distance between the ideal and the reality.

Just as Coleridge's *idea* of the Church was not deducible from the actual state of the Church of England in the early decades of the nineteenth century, so Mill's *idea* of party was not deducible from the actual state of parties in mid-Victorian England. (I do not mean to suggest that party mattered to Mill in the way the Church mattered to Coleridge.) Anyone looking for a guide to the quotidian realities of practical politics in nineteenth-century England should not look to Mill. His concern was for party as it should be, not for party as it was. A high order of aspiration defined Mill's prescriptive conception of party. Many politicians of his own day cared about the public good and believed that political principles counted for something. Most participants in the rough-and-tumble world of mid-Victorian politics, however, were unlikely to suppose that parties contending for power could make principles the coin of the political realm. (In many parliamentary boroughs coin was the coin of the political realm.) Although the Gladstonian Liberal party could not answer the demands Mill made upon it, it did contain within its ranks men who valued in party what Mill valued.[53] Their presence can be taken as one measure of the quality of the party Gladstone led. Perhaps what Mill says of pupils is also applicable to parties. "A pupil from whom nothing is ever demanded which he cannot do, never does all he can."[54]

CHAPTER SEVEN

Mill and the Experience of Political Engagement

None are so illiberal, none so bigoted in their hostility to improvement, none so superstitiously attached to the stupidest and worst of old forms and usages, as the uneducated. None are so unscrupulous, none so eager to clutch at whatever they have not and others have, as the uneducated in the possession of power.

Robert Lowe in 1866? No, J.S. Mill in 1859.[1]

I hope . . . that on Monday next, when the subject comes up again, we shall really get the household suffrage that we want. (Loud cheers.) If we get that we can afford to smile when Mr. Disraeli gets up in an exulting tone—whether we have beaten him or he us, it is all the same to him—he always thinks it his victory—(laughter)—and we can smile when he tells us that we have all come over to him.

John Bright in 1867? No, J.S. Mill in 1867.[2] Context is not everything, but those who wish to argue for a basic consistency and coherence in Mill's politics have to make context count for a lot in reconciling these passages.

Mill acknowledged in the *Autobiography* that his conduct while a Member of Parliament had led some who disapproved to express surprise and bewilderment:

As I had shewn in my political writings that I was aware of the weak points in democratic opinions, some conservatives, it seems, had not been without hopes of finding me an opponent of democracy . . . Yet if they had really read my writings they would have known that after giving full weight to all that appeared to me well grounded in the arguments against democracy, I unhesitatingly decided in its favour, while recommending that it should be accompanied by such institutions as were consistent with its principles and calculated to ward off its inconveniences.[3]

This is a fair synopsis of the position Mill had formulated in his mature political writings, but it does not answer the problem. In the circumstances of 1867 the granting of borough household suffrage was a dramatic step in the direction of democracy. Whatever institutions existed at that time "to ward off its inconveniences," they were not the institutions Mill had recommended in his political essays. Mill would not admit that his practice was at odds with his theory. *The Times*, which made a habit of voicing what others were thinking, believed the disjunction to be very marked indeed.

As a writer it was Mr. Mill's rare fortune to be admired by men of almost all parties; the singular candour which invariably disposed him to consider, not in how much an antagonist might be wrong, but in what little he was right, propitiated the most hostile critics; and yet, strange to say, it is by the very reverse of this quality—by his vehement, narrow partisanship, and apparent inability to see a redeeming point in a political adversary—that Mr. Mill, as a man of action, has estranged many even of his best friends.[4]

These strictures had as much to do with Mill's style as with his substance, and they were prompted by much more than his response to the issue of parliamentary reform. In the *Autobiography* he took up the criticism. Here Mill refers to the accusation brought by "the Tory and Tory-Liberal press" (Mill probably would have allowed *The Times* exclusive occupancy of the latter category) that he had shown himself "in the trials of public life, intemperate and passionate."[5] Although he pled not-guilty, the charge merits an inquiry that goes beyond *The Times*, the *Autobiography*, and the parliamentary career.

The problem of characterizing Mill's conduct as a political activist is not confined to the 1860s. Any investigation of Mill's role as editor of the *London and Westminster Review* during the second half of the 1830s must grapple with it.[6] Mill's commitment to self-culture in the decade-and-a-half following the mental crisis is indisputable. He was determined to expand a Benthamite structure whose walls had closed in on him. Without radically altering the foundation, room had to be made for Wordsworth, Coleridge, Carlyle, and the Saint-Simonians. Analytical rigor, albeit indispensable, could not compensate for a lack of intellectual breadth and poetic sensibility. Mill wanted to have it all, and he earnestly sought to make up for what he perceived as deficiencies in his education. Yet the author of "Bentham" and "Coleridge" was also the editor of a radical journal and the author of articles on the politics of the day that were intended to promote the formation of a radical party capable of challenging the Tories and the Whigs for power.[7]

Joseph Hamburger has persuasively argued that Mill's politics at this time were doctrinaire in character. Convinced that a fundamental conflict existed between the sinister interests of the aristocracy and the general interests of "the people," Mill wished to see this conflict appropriately reflected in the political sphere. The party struggle between Whigs and Tories, aristocratic political forces whose differences were incidental, not only failed to achieve this purpose but was in large part designed to frustrate its achievement. The radical party envisaged by Mill would give forceful and effective expression to the real interests of the people, compel the Whigs to reveal their true colors, and beget a constructive polarization of the political order. In the late 1830s Mill acted on the suppositions that quite drastic political change of a progressive nature could be brought about in a relatively short period of time, and that he had an important part to play in this transformation.[8]

To defend Mill's political strategy and tactics in the 1830s, in light of the absence of correspondence between his vision and the political and social realities it could not comprehend, would be a fruitless endeavor. Instead I want to consider the apparent discordance between his self-conscious pursuit of catholic intellectual sympathies and his equally self-conscious advocacy of a doctrinaire political line. Mill could legitimately claim that he used the *London and Westminster Review* to broaden the outlook of the Philosophic Radicals (or "to educate his party");[9] the fact remains that a dogmatic quality pervades his political journalism of these years. It is the dogmatism of the enthusiast. In Mill's nature divergent needs and tendencies commingled. His profound affinity for reflection, contemplation, abstraction, and analysis cannot obscure his thirst for involvement, participation, and active commitment. Believing that the post-Reform Act political order was susceptible to radical influence, Mill threw himself into the cause with the expectation that a well-organized and vigorously led radical party could do much to advance political and social progress. In his enthusiasm he misjudged the possibilities inherent in the political system, placed a greater burden on George Grote and his fellow Philosophic Radicals in the House of Commons than they had the capacity to bear, and waited in vain for Lord Durham to step forward as leader of the movement.

If in the second half of the 1830s Mill let his zeal master his judgment, psychological insecurity was perhaps partially responsible. Although his mental crisis had passed, the tensions from which it sprung had not been fully purged. He found an outlet for them in his program of creative eclecticism, an eclecticism that denoted growth and ultimately gave Mill the assurance that his mature positions had taken into account all that was relevant to them. Without providing a solution to his difficulties, this program offered a method for arriving at solutions. Instrumental to the achievement of greater wholeness, his open-mindedness was also a symptom of continued disorder. As one who needed to define himself and his role in relation to his times, Mill experienced some confusion and uncertainty during the 1830s. He could not yet be sure of his ability to establish for himself a great name in the speculative world. The political articles in the *London and Westminster*—the public expression of a rather feverish behind-the-scenes private campaign to shake up the political system[10]—probably slowed down his work on the *Logic*. In compensation, they afforded Mill an active and seemingly purposeful release from some of the anxieties that persisted in the aftermath of the mental crisis.

When it became evident to him that the political realignment he had fought to bring about would not occur, Mill did not despair. By 1840 he could handle the disappointment. He had in large measure recovered from the turmoil induced by his father's final illness and death in 1836. Before the close of the 1830s his "friendship" with Harriet Taylor had settled into a manageable, if not entirely satisfactory, routine. Moreover, his progress with the *Logic* had reached the point where he could feel fairly confident about the success and importance of this enterprise.[11] A September 1839 letter from Mill to John Sterling, while assuming a characteristically diffident tone, shows a man more sure of himself than had earlier been the case: "I quite think with you that it is no part of my vocation to be a party leader, but at most to give occasional advice to such as are fitted to be so. Whether I have any better vocation for being a philosopher, or whether you will think so when you see what I am capable of performing in that line, remains for the future to decide."[12]

Mill's organization of the *Autobiography* indicates that the fundamentals of his thought were settled by 1840. Six of the seven chapters pertain to the pre-1840 period. He opens the seventh, entitled "General View of the Remainder of My Life," with the following statement: "From this time [1840], what is worth relating of my life will come into a very small compass; for I have no further mental changes to tell of, but only, as I hope, a continued mental progress; which does not admit of a consecutive history, and the results of which, if real, will be best found in my writings."[13] In the two decades after 1840 Mill usually maintained a general aloofness from the party political scene. He continued to interest himself in what was going on at Westminster, but he did so as an observer with strong opinions rather than as a participant seeking to influence developments in a direct fashion. This aloofness arose in part from Mill's perception of a political environment inhospitable to the sort of pressure he had ineffectually attempted to apply in the 1830s. The perception itself, however, was colored by Mill's altered sense of his own function and purpose within the culture of his age.

The decisive shift came in 1839–40. His withdrawal from the *London and Westminster* in the latter year signaled Mill's disillusionment with politics. It no less presaged a new era in Mill's life, one in which he would invest most of his intellectual capital in the speculative market. The contrast between his 1839 article "Reorganization of the Reform Party" and his 1840 essay on Tocqueville intimates the change in disposition. Appearing in the April 1839 issue of the *London and Westminster*, "Reorganization of the Reform Party" represents Mill's culminating effort to give radicalism both the social analysis and the practical program he believed it required to become an independent, organized, and effective political force. His division of society into "natural Radicals" and "natural Conservatives," the former greatly exceeding the latter in number, was sanctioned by a somewhat mechanical assessment of the concrete interests of various groups and classes as defined by their location in the social structure.[14] Mill emphatically stated that radicalism sufficiently well understood the ends of political action; it had yet to devise the means to reach those ends. "Radicalism has done enough in speculation; its business now is to make itself practical. Most reformers are tolerably well aware of their ends; let them turn to what they have hitherto far less attended to—how to attain them."[15] The essay on Tocqueville does not neglect the formative influence of social structure. It nonetheless reveals a frame of mind noticeably different from that evinced in "Reorganization of the Reform Party." In the 1840 essay he asserts: "Economical and social changes, though among the greatest, are not the only forces which shape the course of our species; ideas are not always the mere signs and effects of social circumstances, they are themselves a power in history."[16]

By 1841 a virtually complete reversal of the 1839 perspective is evident. Writing to Macvey Napier in July of that year, Mill declared: "We are entering upon times in which the progress of liberal opinions will again, as formerly, depend upon what is *said & written*, & no longer upon what is *done*, by their avowed friends."[17] A year later he told Robert Barclay Fox that "There never was a time when ideas went for more in human affairs than they do now."[18] What had been only hinted at in the Tocqueville article received full-blown expression in 1843, with the publication of *A System of Logic*. Here Mill affirms that "the state of the speculative faculties, the character of the propositions assented to by the intellect, essentially determines the moral and political state of the community."[19] Integrating this maxim into his treatment of the "Historical Method," Mill says that "the order of human progression in all respects will mainly depend on the order of progression in the intellectual convictions of mankind, that is, on the law of the successive transformations of human opinions."[20]

Mill saw the *Logic* and the *Principles of Political Economy* (1848) as contributions to this transformational process. Their impact on the mental culture of the age was indeed undeniable, and their success transformed Mill's status within that culture. Although these works did not make Mill the great public figure he would become after 1859, they did establish him as one of the most important thinkers of his time.[21] The scope, solemnity, and analytical power of the *Logic* could not fail to make a deep impression. Its technical difficulty doubtless limited its readership, but this limitation was inherent in the subject matter, over which Mill had exercised an imposing mastery. The *Principles of Political Economy* gained a commanding presence in its field, and had the additional merit of being accessible to a somewhat wider audience. Together, they conferred upon Mill a distinction in the world of thought that seemed to vindicate his concentration on the theoretical to the exclusion of the narrowly political.

Of course almost everything Mill wrote could be construed as in some sense political. He always meant for his ideas to count for something in "human affairs." Politics—the clash of interests, the configuration of the state, the creation of policy—claimed an important place in Mill's conception of "human affairs." Yet his writings of the 1840s, with few exceptions, had little in common with those of the late 1830s that had reflected a fascination with party politics and a preoccupation with the fortunes of radicalism. The latter appear in volume six of Mill's *Collected Works*, entitled *Essays on England, Ireland, and the Empire*. Only two items in this volume fall after 1839, and these focus on the Irish question.[22] As for the *Essays on Politics and Society*, volumes eighteen and nineteen of the *Collected Works*, there is nothing between the second review of Tocqueville (1840) and the essay *On Liberty* (1859), apart from a succinct 1854 paper on the reform of the civil service.[23] In the sphere of speculation, Mill's accomplishments in the 1840s were those of a logician and political economist, not those of a political philosopher.

Mill did not abstain altogether from political journalism during the 1840s. From October 1846 to January 1847 he wrote forty-three leading articles for the *Morning Chronicle* on the condition of Ireland.[24] The Irish famine occurred while he was at work on the *Principles of Political Economy*. That calamity led him to consider the importance of such "principles" in relation to an urgent practical problem of exceptional magnitude. He called upon the government to implement a scheme for the creation of peasant proprietorships on the waste lands of Ireland.[25] Acting on the belief that a singular opportunity had arisen for reconstructing the Irish economy with a view to long-term social and moral improvement, Mill vigorously entreated English politicians to grasp the Irish nettle and show themselves statesmen.

He would not have bothered to do so had he not thought the political system capable of a positive response to such an appeal. Distance from the scene of action during the first half of the 1840s had dulled the critical edge of Mill's political temper. In March 1842 he informed Sarah Austin that "Politics here are going smoothly enough. Peel is making a considerable number of petty improvements, such however as would not have been thought petty formerly."[26] Six months later he told Robert Barclay Fox that though he had "almost given up thinking on the subject" of politics, he did

> believe that ever since the changes in the Constitution made by Catholic emancipation and the Reform Act, a considerable portion of the ruling class in this country, especially of the younger men, have been having their minds gradually opened, & the progress of Chartism is I think creating an impression that rulers are bound both in duty & prudence to take more charge, than they have lately been wont to do, of the interests both temporal & spiritual of the poor.[27]

Two more years of Peel's administration did nothing to sour Mill's political disposition. Writing to Henry S. Chapman in November 1844, Mill stated: "I fully expect every session to shew concessions to liberalism, and every year certainly helps to disorganize the old order."[28]

In 1846–47 Mill seriously entertained the notion that a British government could do what was necessary for Ireland. When Lord John Russell's ministry declined to act upon his recommendations he was deeply dismayed. The account of this episode in the *Autobiography* explains the government's disregard of his proposal in the following terms: "the idea was new and strange; there was no English precedent for such a proceeding: and the profound ignorance of English politicians and the English public concerning all social phenomena not generally met with in England (however common elsewhere) made my endeavours an entire failure."[29] Mild is the flavor of resentment presented here as compared with what can be tasted in letters he wrote in 1847. In March of that year, two months after his final contribution to the *Morning Chronicle* on the subject, Mill told Chapman that he had "never felt so thoroughly disgusted with the state of public affairs . . . Ireland will be in a state next year that will make the landlords sell the clothes off their backs to get rid of the people."[30] A month later John Austin received a lengthy disquisition from Mill on the crippling inadequacies of the English aristocracy as a governing class. Noting that he could see advantages accruing from "a violent revolution" in England, Mill declared: "England has never had any general break-up of old associations & hence the extreme difficulty of getting any ideas into its stupid head."[31] This frame of mind, precipitated by the failure of his *Morning Chronicle* campaign, found further expression several weeks later in a letter to Alexander Bain: "The conduct of the ministers is wretched beyond measure upon all subjects; nothing but the meanest truckling at a time when a man with a decided opinion could carry almost anything triumphantly."[32]

Mill's use of the press in 1846–47 to propagate his views on a central question of public policy had no connection with political motives of a party nature, which makes it unlike his mission of the late 1830s. Yet the experience markedly affected his perception of the state of English politics. Once again he had tried to exert political influence; again he had been thwarted. His frustration may have had some bearing on his reception of the French Revolution of 1848 and on his discussion of British parliamentary developments in this year.

Mill's radical political sympathies survived intact, despite his withdrawal from an active political role in the 1840s. In 1839 he had acknowledged that universal suffrage could not "be either good or practicable now."[33] He had, however, favored a very generous extension of the franchise in the 1830s—commonly expressed in the form of "household suffrage"—and he did not retreat from this position over the next decade. In 1842 he wrote a long and highly complimentary letter to William Lovett, the London Chartist leader and head of the National Association for Promoting the Political and Social Improvement of the People. Mill explained that he could not endorse universal suffrage, "in the present state of civilization at least." He would give the working classes "the choice of only a part, though a large & possibly progressively increasing part of the legislature but that part you should elect conformably to all the six points of the Charter & I should object as much as any of you to surrendering one iota of them."[34] Few English non-working-class radicals would have subscribed to a platform more democratic than this in the 1840s. Indeed, in conventional radical terms the position outlined to Lovett was more advanced than that espoused by Mill in either *Thoughts on Parliamentary Reform* (1859) or *Considerations on Representative Government* (1861). Thus the events of 1848 would activate a political radicalism that had lain dormant for much of the 1840s without ever having been abandoned.

Unlike most respectable Englishmen, who congratulated themselves on the contrast between the fizzling out of Chartism in England in 1848 and the revolutionary upheavals in France of the same year, Mill was inspired by the example of radicals across the Channel acting to create a new order founded on popular institutions. His zeal for improvement was such that he could not help admiring those prepared to risk all to accomplish a giant leap in the right direction when there was no reasonable prospect for incremental reform. Mill took up the cause in the face of British detractors, both in a *Daily News* article of August 1848 and in a lengthy 1849 essay, "Vindication of the French Revolution of February 1848."[35] Whether or not Mill understood this revolution any better than he understood English politics in the 1830s is not the issue here.[36] The important point is that the French Revolution of 1848, like those of 1789 and 1830, animated Mill. His slumbering radical political spirit was suddenly wakened by the spectacular developments in France in 1848.

The *Daily News* article on "French Affairs" was tied in with several leaders Mill wrote for this advanced Liberal paper in the summer of 1848. Commenting on the hostility of both Whig and Tory papers to the revolution, Mill proclaimed: "There is no way now of discrediting reform without blackening France. The enemies of popular institutions have lost their most potent weapon, fear of the unknown. Democracy, in the popular signification of the term, exists as a fact, among our nearest neighbours."[37] One might well wonder why opponents of democracy in England, having recently witnessed the collapse of Chartism, should trouble themselves about discrediting reform. The answer is that the defeat of Chartism did not necessarily signal a setback for the sort of reform favored by non-working-class radicals.[38] Quite the contrary. The emergence of "physical force" Chartism had made reform unsafe in the eyes of propertied Englishmen. With the crumbling of Chartism, a revival of the reform cause under respectable auspices could be undertaken. In the spring of 1848 between fifty and sixty radicals in the House of Commons, led by Richard Cobden and Joseph Hume, carried their disaffection with Lord John Russell's Whig government so far as to form a distinct party with a program of parliamentary reform and governmental retrenchment. In July Hume brought forward his "Little Charter," a scheme that included household suffrage, the secret ballot, triennial parliaments, and a greater equality of electoral districts. A serious debate ensued in the Commons. These developments heartened Mill, who contributed three leaders on parliamentary reform to the *Daily News* in July 1848.[39] These tell us a good deal about the condition of his politics at this time.

Mill treats Hume's initiative as the programmatic centerpiece of a radical campaign whose ultimate ambitions were far reaching. Hume's proposal, according to Mill, pitted supporters of a "large reform" against those opposed to further constitutional change. Here Mill does not employ the terms "natural Radical" and "natural Conservative," but he applauds Hume's scheme for its tendency to bring into focus the divide between the political forces examined in "Reorganization of the Reform Party." These articles demonstrate that Mill's practical political preferences had not altered since the publication of that essay. His analytical framework, however, had changed. The *Daily News* leaders make no attempt to link political conflict and social structure. From this we should not infer that Mill saw no connection. The "reform party" had disintegrated rather than reorganized at the end of the 1830s, which implied that Mill's understanding of the problem had been flawed. The relation of political conflict to social structure was neither so simple nor so direct as he had suggested in 1839. When faced with the Chartist threat, many within the middle classes had shown themselves to be "natural Conservatives." The time had come for the middle-class radicals to make the running. The reorganization of the reform party in 1848, as projected by Mill in his *Daily News* articles, would be grounded on principles of improvement possessing a power and legitimacy that transcended social divisions.

Mill urges the reformers in the House of Commons to give to the contest at hand the dignity appropriate to "a serious conflict of opposing reasons":

> Their practical conduct as politicians necessarily partakes of compromise. Their demands and systematic aims must often fall short of their principles. But let them not therefore cut down their principles to the measure of their demands. If they do, they lose far more in vigorous argument, and in the imposing influence of a sense of consistency and power, than they can possibly gain in charming away the fears of those who would, but dare not, follow them. Let them disclaim nothing which is a consequence of their principles.[40]

That the reformers stood in need of such instruction, Mill did not doubt. They had yet to learn "to make great principles their object, and in this lies the secret . . . of ineffectual struggles and mean results. The world will rally round a truly great principle, and be as much the better for the contest as for the attainment; but the petty objects by the pursuit of which no principle is asserted, are fruitless even when attained."[41]

Prefigured in these 1848 articles were the arguments and objectives Mill would set forth in his later writings on parliamentary reform. An electorate that incorporated the working classes, Mill says, would give rise to a legislature attentive to the needs, interests, and concerns of the masses. In those cases where the legislature concluded that the "apparent" interests of the workings classes diverged from the "real" interests of the nation, its members would feel obliged to defend their policy "by reasons drawn from the interests of those same classes, and appealing to their understandings." The result would be a thorough airing of the issues in Parliament and in the press that would serve to invigorate the mental faculties of the governing classes and to stimulate those of the mass electorate. Strenuous efforts would have to be made by the former to educate the latter. Representation of the working classes in the House of Commons would markedly enhance the quality of the nation's political life. Lacking such representation, the legislature must inevitably fail to perform one of its primary functions—that of giving constructive political expression to the "adverse interests and principles" dividing society. Mill conceived of Parliament as "the arena where opposing forces should meet and fight out their battle, that they may not find themselves reduced to fight it in a less pacific field." A properly constituted House of Commons would not only embody the "struggle between conflicting principles"; it would also do much to bring those principles "into just relation with one another." The blindness of the propertied classes to the legitimate claims of those without property could not be overcome until the disadvantaged were admitted into the House of the privileged, who had "to learn the difficult but necessary art of looking at established institutions and opinions from the point of view of those who are not on the sunny but the shady side of the social edifice. Defects by which other people alone suffer are seldom seen until the sufferers point them out."[42]

Before the end of the 1850s Mill would join these goals to a program of reform that would look rather different from the one he sponsored in the 1830s and reaffirmed in 1848. The articles in the *Daily News* indicate that the conventional radical planks making up Hume's "Little Charter" found in Mill a ready supporter. He argues that "even in a legislature chosen by universal suffrage" the working classes would not "have at all more than their just influence." A democratic suffrage, Mill insists, would fundamentally change the conditions of government, and the change would be salutary: "Whoever may be the rulers, the interest of the great mass of the community must then stand foremost among the activating principles in the conduct of public affairs."[43] There is nothing in these articles about an educational qualification for the suffrage, the representation of minorities, or the virtues of open voting. If Mill's plea in 1848 for a large dose of parliamentary reform raised the level of argumentation beyond what most radicals could reach (not that many aspired to such heights), the content of his platform did not noticeably deviate from orthodox democratic radicalism. The prospectus on reform Mill fashioned over the next decade did amount to a significant deviation from radical orthodoxy.

Mill's political enthusiasm had been rekindled by the events of 1848. The republican experiment in France engaged his sympathies and made him highly responsive to any sign of political movement in England. By the middle of 1849 he could find none worthy of a response. The feebleness of the parliamentary radicals was palpable. Unable to agree on a leader or a policy, they had no chance of mounting any sort of sustained challenge to aristocratic hegemony. On the Continent, meanwhile, reaction had already begun to set in. Mill had plenty of intellectual resilience; with political resilience he was less well endowed. For a short time Mill had again felt the pull of politics as the human agency capable of giving a rapid acceleration to the rate of progressive change. Once more he felt let down, and once more he withdrew from the scene. In late May 1849 he wrote to H.S. Chapman: "As for England, it is dead, vapid, left quite behind by all the questions now rising. From the Dukes to the Chartists, including all intermediate stages, people have neither heads nor hearts."[44] Eight months later he took up a stance reminiscent of the early 1840s: "We have come, I think, to a period, when progress, even of a political kind, is coming to a halt, by reason of the low intellectual and moral state of all classes: of the rich as much as of the poorer classes. Great improvements in education . . . are the only thing to which I should look for permanent good."[45] Mill has this period in mind when he writes in the *Autobiography* of his having "learnt from experience that many false opinions may be exchanged for true ones, without in the least altering the habits of mind of which false opinions are the result." Although the "English public" might "have thrown off certain errors, the general discipline of their minds, intellectually and morally, is not altered. I am now [he is writing in late 1853 or early 1854] convinced, that no great improvements in the lot of mankind are possible, until a great change takes place in the fundamental constitution of their modes of thought."[46]

From 1849 through 1858 the English public did not hear much from Mill. He penned a few occasional pieces in the early 1850s, none of which was intended to cause even a ripple on the surface of political life. Apart from a pamphlet on the reform of the civil service, miscellaneous contributions to newspapers, and what he put into print in defense of the East India Company, he published nothing new on any subject, political or otherwise, in the five years after 1853. Mill and Harriet Taylor married in 1851 and lived a rather solitary existence until her death in 1858. Both suffered from poor health during this period; both were disdainful of "society."[47] Of course they were not idle. The first draft of the *Autobiography* was composed in late 1853 and early 1854, and *On Liberty* also took shape prior to Harriet's death. Together they continuously reviewed their agenda for the improvement of mankind. The matrimonial years further removed Mill from the world of practical political activism. His retreat from the political world was by no means total. In expectation of the Aberdeen Coalition's 1854 Reform Bill he wrote most of what was published in 1859 as *Thoughts on Parliamentary Reform*. The lofty radicalism privately cultivated by the Mills—a radicalism whose aim was nothing less than "to regenerate society"—denoted a reappraisal of the old democratic platform with which Mill had in the main identified and which they now found wanting: "We were now much less democrats than I had been, because so long as education continues to be so wretchedly imperfect, we dreaded the ignorance and especially the selfishness and brutality of the mass."[48] When the "I" became a "We," the politically democratic strain in Mill was overtaken by the combined force of the intellectually aristocratic strain present in the husband and predominant in the wife. It is remarkable that the author of the 1848 *Daily News* leaders discussed above should have written the following in a journal he kept during the first several months of 1854: "democratic institutions seem likely enough to be conceded, and that, too, more rapidly than is desirable, by the almost unasked liberality of the better part of the aristocracy."[49] The Mill who emerged from semi-seclusion in 1859 was a radical of a different stripe from the Mill who had embraced Hume's "Little Charter" in 1848.

Mill's life entered a new phase in 1859. The previous year had seen Harriet's death and Mill's retirement (with a handsome pension) from the East India Company.[50] Mill found solace in the companionship of his stepdaughter Helen Taylor and in new friendships based primarily on shared political convictions.[51] He found occupation in a flurry of publication activity. In February 1859 *On Liberty* was released. Almost instantaneously the eminent Victorian thinker acquired the status of influential public moralist.[52] Also in February, the pamphlet *Thoughts on Parliamentary Reform* was published, to be followed in April by "Recent Writers on Reform" (*Fraser's Magazine*) and by *Dissertations and Discussions*.[53] Before the year was out, "Bain's Psychology" (*Edinburgh Review*) and "A Few Words on Non-Intervention" (*Fraser's Magazine*) were in print. The renewed interest in political matters, parliamentary reform in particular, is obvious. Furthermore, this interest largely defined the parameters of a rejuvenated social life that would center on Mill's associations with Henry Fawcett, Thomas Hare, and J.E. Cairnes.[54]

Mill's changed situation and desire for political involvement, rather than the emergence of a fresh political climate receptive to radical innovations, led him to air his views on parliamentary reform. Some curious happenings at the top did affect the contours of British politics in 1859. The Tory minority government of Derby and Disraeli introduced a modest Reform Bill; Palmerston and Russell patched up their personal differences in the interest of a common front, without which they could not hope to secure a sizeable and steady majority in the House of Commons and thereby displace Derby and Disraeli; Gladstone completed his tortuous peregrination from the Conservative to the Liberal party by agreeing to serve as Chancellor of the Exchequer under Palmerston after the fall of Derby's ministry; and John Bright mildly compromised his independence by urging radicals to support Palmerston and Russell in exchange for a measure of parliamentary reform. None of this maneuvering among the big guns, however, was calculated to fire Mill's political imagination. And there would not be much rumbling from below. While the politicians sparred and danced, the great British public kept up an impressive display of indifference. The tone of *Thoughts on Parliamentary Reform*, "Recent Writers on Reform," and *Considerations on Representative Government* was in part the product of the age of Palmerston in British politics.

Most of *Thoughts on Parliamentary Reform* had been written five years before its publication. British entry into the Crimean War relegated parliamentary reform to the back burner, and Mill decided to bide his time. In early 1859 it became clear that the Tory ministry would make a move on parliamentary reform. Thinking the moment right, Mill now disclosed a position on the subject that gave several twists to the traditional radical program.

In *Thoughts on Parliamentary Reform* Mill recommends adult suffrage limited by an educational qualification, the abolition of small constituencies, the representation of minorities, and plural voting. He devotes nearly one-third of the pamphlet to an argument *against* secret voting. Most radicals did not care for an educational qualification, were opposed to plural voting, and favored the secret ballot. There is nothing inherently undemocratic about open voting, and Mill's argument against the ballot, albeit interesting, need not detain us here.[55] In theory the representation of minorities in proportion to their numbers does not offend democratic tenets. The educational qualification might in practice restrict working-class electoral power for a time, but Mill did not see it as an insuperable obstacle to virtual adult suffrage being achieved in the foreseeable future. He presents a powerful brief for the justice and necessity of adult suffrage in principle, and posits that an association between the suffrage and a minimum educational standard will do much for both education and politics.[56] Many radicals might have been dubious about amalgamating an educational qualification, representation of minorities, and open voting; they had no use whatever for plural voting.

The pamphlet as originally drafted did not include a defense of plural voting. Mill's *Autobiography* tells us that he added this feature in 1859, and that he did so without ever having consulted Harriet on the subject.[57] In a letter of December 1857 Mill cited an educational test, together with the representation of minorities via a scheme that would permit a voter to give both his votes to a single candidate[58]—most constituencies returned two members to the House of Commons—as the "only provisions for increasing the influence of the more educated class of voters, to which I see my way." Not yet, it seems, had he convinced himself that plural voting was an expedient device for augmenting the electoral weight of the educated. Additional votes might be allotted "to a limited number of what are called liberal professions, on the presumption (often a very false one) that every member of these professions must be an educated person. But nearly all the recognized professions have as such, interests & partialities opposed to the public good."[59] Mill inscrutably changed his mind on plural voting between the end of 1857 and the publication of *Thoughts on Parliamentary Reform*. Had Harriet still been around in 1859, she would perhaps have objected to giving any man multiple votes while no woman had even a single vote.

Mill, it is true, consistently maintained that the introduction of plural voting would be apposite only to an adult suffrage virtually universal in extent. If some adults were to have more than one vote, all should have at least one. And he emphatically rejected the proposition that property should be the rule for awarding plural votes. Education ("knowledge and cultivation") was to be the gauge of fitness. How was this fitness to be determined?

> If every ordinary unskilled labourer had one vote, a skilled labourer, whose occupation requires an exercised mind and a knowledge of some of the laws of external nature, ought to have two. A foreman, or superintendent of labour, whose occupation requires something more of general culture, and some moral as well as intellectual qualities, should perhaps have three. A farmer, manufacturer, or trader, who requires a still larger range of ideas and knowledge, and the power of guiding and attending to a great number of various operations at once, should have three or four. A member of any profession requiring a long, accurate, and systematic mental cultivation . . . ought to have five or six. A graduate of any university, or a person freely elected a member of any learned society, is entitled to at least as many.[60]

The distribution of property in mid-Victorian England did not correspond precisely with the distribution of education. Yet Mill's measure of electoral value would not have short-changed the propertied. (Harriet, had she been alive, might have pointed out to him that such a measure, even when accompanied by adult suffrage, would have seriously disadvantaged women.) The electoral hierarchy Mill delineates bears a fairly close resemblance to the social and economic hierarchy, one in which property played a prominent part (landed property had ways of making its influence felt independently of plural voting). An unskilled laborer contemplating Mill's scheme might wonder whether his interests could be well served by such an arrangement. Mill wanted those interests well served. In *Representative Government* he stipulated that "The plurality of votes must on no account be carried so far, that those who are privileged by it, or the class (if any) to which they belong, shall outweigh by means of it all the rest of the community."[61] What this means in practice is difficult to say. Mill wished to ensure that the votes of educated people put a formidable imprint on the political order without giving undue influence to the economic or class interests embedded in these individuals. Radicals of a less theoretical bent could take comfort in the thought that this distinguished radical philosopher would never be prime minister.

In the *Autobiography* Mill is somewhat defensive about his advocacy of plural voting. He says that Tory perceptions of him, before he entered Parliament, as an opponent of democracy "appear to have been founded on the approbation I had expressed of plural voting, under certain conditions . . . I had made it an express condition that the privilege of a plurality of votes should be annexed to education, not to property, and even so, had approved of it only on the supposition of universal suffrage."[62] I have suggested that the distinction between property and education bordered on the specious. I would also argue that the statement regarding plural voting in the *Autobiography* does scant justice to the conviction that girded Mill's allegiance to plural voting when he wrote *Representative Government*. He had put forward plural voting not as a necessary evil, but as a positive good:

> I do not propose plurality as a thing in itself undesirable . . . I do not look upon equal voting as among the things which are good in themselves . . . It is not useful, but hurtful, that the constitution of the country should declare ignorance to be entitled to as much political power as knowledge. The national institutions should place all things that they are concerned with, before the mind of the citizen in the light in which it is for his good that he should regard them: and as it for his good that he should think that every one is entitled to some influence, but the better and wiser to more than others, it is important that this conviction should be professed by the State, and embodied in the national institutions.[63]

To be sure, Mill was a thinker who demanded to be read with greater care than most Tories would bring to the task. They had nonetheless been given cause to associate plural voting with resistance to what they understood as democracy, and to consider Mill a champion of plural voting.

Mill's panegyric on plural voting in *Representative Government* was restrained by comparison with his commendation of personal representation in the pivotal chapter of this book. After the publication of *Thoughts on Parliamentary Reform*, Mill read Thomas Hare's *A Treatise on the Election of Representatives, Parliamentary and Municipal* (1859), which he reviewed in "Recent Writers on Reform."[64] Reduced to its simplest terms, Hare's scheme of personal representation required voters to mark their ballots in order of preference. The number of valid ballots cast would then be divided by the number of seats in the House of Commons to determine the quota necessary for election. If a candidate received more than the necessary quota, his surplus votes would be distributed among the second choices and so on until all the seats were filled. A candidate not reaching the quota from ballots recorded in his own constituency could also draw upon ballots from other constituencies that showed preferences for him. Thus national as well as local constituencies would, in effect, elect candidates. Hare's plan gave Mill just what he had been looking for. Remarking on Hare's book to George Cornewall Lewis in March 1859, Mill wrote: "I think it both a monument of intellect, and of inestimable practical importance at the present moment . . . Had I seen this book before writing my pamphlet I should have made it very different."[65] To Hare himself, in December of this year, Mill declared: "The more I think of your plan, the more it appears to me to be *the* great discovery in representative government."[66] In "Recent Writers on Reform," Mill referred to Hare's treatise as "the most important work ever written on the practical part of the subject."[67]

Mill expected great things from the implementation of Hare's proposal. The existing electoral system, dominated by petty party interests and local political networks, prevented men of ability and independent mind from obtaining seats in the House of Commons. Indeed, conditions were such that individuals possessing these qualities normally did not come forward as candidates. Given the prospect of drawing votes from what in essence would be a national constituency, the best men would stand for election and probably win. Each Member of Parliament would represent a unanimous constituency rather than a divided one whose electorate had distributed their votes among several candidates, some of whom had been defeated. The relation between representative and elector would be founded on shared conviction and mutual respect. The legitimate influence of principles would replace the illegitimate influence of money at elections.[68]

Hare's scheme persuaded Mill that the widest possible political participation could be coupled with a high level of political competence.[69] Universal suffrage, even without plural voting, held no terrors for Mill, provided that the electoral system guaranteed the educated minority a representation in the House of Commons proportionate to their numbers in the country. "The causes of the minority would be likely to be supported with such consummate skill, and such a weight of moral authority as might prove a sufficient balance to the superiority of numbers on the other side, and enable the opinions of the higher and middle classes to prevail when they were right, even in an assembly of which the majority had been chosen by the poor."[70] It may not be too much to say that Hare's plan made possible the writing of *Representative Government*. It is no accident that Mill's treatment of Hare's system in this book occurs in the chapter entitled "Of True and False Democracy; Representation of All, and Representation of the Majority Only."[71] Mill had been searching for the means to establish "True Democracy," and Hare, so Mill thought, had furnished those means. He was not slow to acknowledge his debt to Hare:

> As you have read the two volumes of Dissertations [*Dissertations and Discussions*], you have seen how during a great part of my life I have been troubled by the difficulty of reconciling democratic institutions with the maintenance of a great social support for dissentient opinions. Now your plan distinctly solves this difficulty. The portion of the House of Commons returned by an union of minorities would *be* this social support, in its most effective form; since its members would meet in the same arena with the organs of the majority; would command public attention . . . and would have the opportunity of obtruding upon the public daily proofs of the superiority of individual value which they would generally have over their antagonists. In no other way that I can conceive, would it be possible to maintain a real superiority of power in the majority, along with a full & fair hearing for minorities, and an organization of them which would be all the more effective from being natural and spontaneous.[72]

Radicals more experienced and less cerebral than Mill and Hare, notably Bright and Cobden, were generally unsympathetic to the representation of minorities. Cobden's preference was for first-past-the-post single-member constituencies of approximately equal size (10,000 electors per constituency seemed reasonable to Cobden): "I don't know any better plan for giving all opinions a chance of being heard, and, after all, it is opinions that are to be represented. If a minority have a faith that their opinions, and not those of the majority, are the true ones, then let them agitate and discuss until their principles are in the ascendant."[73]

Mill's alliance with Hare accentuated the differences between him and most of the radicals in Parliament. Mill believed that Hare's plan could make representative government work in the public interest. The man who authored this plan, however, was no sort of democrat. Hare's volume included many lengthy quotations, reverently treated, from Burke, Calhoun, Guizot, and Blackstone.[74] Mill himself demurs from Hare's highly conservative views on the franchise.[75] Hare recommended that the £10 household qualification should be retained for the large towns, and that the qualification in small towns and counties should range from £6 to £10, variations corresponding to regional differences in the cost of housing. Such proposals could be expected to find a more receptive audience among Tories than among radicals.

These same Tories could easily light upon passages in *Representative Government* that would meet with their approval. "The natural tendency of representative government, as of modern civilization, is towards collective mediocrity: and this tendency is increased by all reductions and extensions of the franchise, their effect being to place the principal power in the hands of classes more and more below the highest level of instruction in the community."[76] With a large extension of the franchise, "the great majority of voters . . . would be manual labourers; and the twofold danger, that of too low a standard of political intelligence, and that of class legislation, would still exist, in a very perilous degree."[77] "Until there shall have been devised, and until opinion is willing to accept, some mode of plural voting which may assign to education, as such, the degree of superior influence due to it, and sufficient as a counterpoise to the numerical weight of the least educated class; for so long, the benefits of completely universal suffrage cannot be obtained without bringing with them, as it appears to me, more than equivalent evils."[78]

Viewed in isolation, these assertions, though faithfully conveying Mill's hostility to forms of majoritarian democracy that failed to safeguard the valid interests of minorities, give a distorted impression of the disposition that structures the analysis and argument of *Representative Government*. Had Mill seen the problems raised by the franchise question as insurmountable barriers to the creation of a broad, comprehensive, and effective democratic system, he never would have written *Representative Government*. The primary function of Mill's major political essay was to advance the rationale for such a system and show how it could be constructed. No doubt the tension evident in some of the analysis and argument threatens to undermine the coherence and integrity of the theory. It could scarcely have been otherwise for a thinker whose purpose was to discover and disclose the participatory institutions and mechanisms that would meliorate the "ignorance and . . . selfishness and brutality of the mass."[79]

Although Mill's readership as of 1859 extended into the upper ranks of the working classes, *Thoughts on Parliamentary Reform*, "Recent Writers on Reform," and *Considerations on Representative Government* were addressed to the educated political public of mid-Victorian England. The tone of these writings is judicious, restrained, balanced. Radical elements certainly show up in the content, but the presentation is never menacing or demagogic. Mill's appeal is to the reason of those with power, not to the passions of those without. Shaping and circumscribing Mill's strategy were his awareness of the quiescence of the excluded, and his recognition that the absence of a redoubtable agitation since the repeal of the Corn Laws had not crippled the march of reform. Free trade had triumphed—the repeal of the Navigation Acts and the success of Gladstone's budgets writing the final chapters in the book that had begun with the commercial and fiscal reforms of Liverpool, Huskisson, and Peel. The "taxes on knowledge" had been either eliminated (the newspaper tax in 1854, the advertising tax in 1855) or were about to be swept away (the paper duties in 1861). A much needed reform of the civil service and ancient universities had been initiated, if not completed. A great deal remained to be accomplished, but Mill was prepared to concede that the existing political system had the capacity to answer the call of improvement. A different conclusion, under the circumstances, would have rendered pointless the labor he invested in the political essays of this period.

Mill does not pretend to his readers or to himself that a radicalization of the political environment was at all likely. At the beginning of *Thoughts on Parliamentary Reform*, he observes that the issue had "not been pressed upon the ruling powers by impetuous and formidable demonstrations of public sentiment, nor preceded by signs of widespread discontent with the working of the existing political institutions." A second installment of parliamentary reform will come, he surmises, "without having required, or occasioned, any unusual amount of peaceful agitation." And he pronounces this to be a cause for satisfaction, declaring that "the mustering and trial of strength between the Progressive and Stationary forces which filled the fifteen years from 1832 to 1846, have inaugurated Improvement as the general law of public affairs."[80]

Mill's public engagement with the mid-Victorian political system in the late 1850s and early 1860s was more of the head than of the heart. Spawned by his private needs following Harriet's death, by his heightened public profile after the publication of *On Liberty*, and by a moderate revival of interest in parliamentary reform among the politicians, this engagement brought forth Mill the political philosopher with a preferred public agenda, not Mill the political activist. Between 1865 and 1868 Mill the political activist would set the pace.

In the few years preceding his election to the House of Commons, Mill did not make politics his chief concern. The American Civil War moved him to write two impassioned essays in support of the North, but no other development on the political front arrested his attention.[81] Westminster politics in the first half of the 1860s featured an ancient, spry, atavistic Regency figure as prime minister (so Palmerston must have seemed to Mill), a conspicuously complacent House of Commons, and a generally lethargic public. The anticipated second coming of parliamentary reform in 1859–60 did not materialize, and whenever (not often) the drowsy dogs of reform showed signs of stirring in the five years thereafter, Palmerston deftly lulled them back to sleep. There was little to distract Mill from his massive exposition and criticism of Sir William Hamilton's philosophy (published in 1865). For Mill, this overblown attack on a dead thinker he had come to consider an especially pernicious proponent of Intuitionism no doubt carried a political charge. Its impact did not disturb the habitat in which Palmerston and company dwelt.

Mill's election to the House of Commons followed his publication of *An Examination of Sir William Hamilton's Philosophy*. When Mill took his seat in the House in early 1866, he positioned himself in relation to a political struggle that to him appeared fraught with weighty implications for the future course of British politics. The prospect of a radicalization of the Liberal party, embracing Gladstonian leadership and a dynamic working-class component, molded Mill's political actions both inside and outside the Commons.[82] Measured against the laxity of Palmerston's nominal liberalism, Gladstone's earnest character and conduct, Mill believed, merited high esteem.[83] Mill warmed to Gladstone's intellectual power, his administrative capacity, his resolute attachment to reform and improvement. Palmerston's passing (in October 1865) and Russell's advanced age pointed to an imminent Gladstonian succession, to which Mill eagerly looked forward. Viewing Gladstone as the indispensable apex of a radicalized Liberal party, Mill in his parliamentary years tended to cut his political cloth to fit Gladstone's needs.[84]

Mill saw in Gladstone the single Liberal politician of stature capable of integrating the working classes into the political process. The formation of the artisan-dominated Reform League in 1865, and its mobilization of working-class activism in 1866 and 1867, helped end the torpor of the Palmerstonian era. Mill's gaze fixed upon the prospect of rapid political progress.

From 1866 through 1868 Mill reached out to the working-class political movement. Dissenting from the program of the Reform League—the League favored manhood suffrage and the ballot while Mill wanted adult suffrage and public voting—he nonetheless identified himself with its struggle. In declining the invitation to become a member of the League, Mill remarked that "the general promotion of the Reform cause is the main point at present, and . . . advanced reformers, without suppressing their opinions on points on which they may still differ, should act together as one in the common cause."[85] In the midst of the July 1866 Hyde Park riots Mill defended the League in the House of Commons and subsequently sent in a £5 donation to those arrested by the police.[86] At the end of July he addressed a Reform League meeting at the Agricultural Hall.[87] In February 1867 he participated in a deputation whose purpose was to persuade Spencer Walpole, the Tory Home Secretary, to appoint a working man to the Royal Commission on Trades Unions.[88] In the summer of that year Mill subscribed to a Reform League fund for organizing the newly enfranchised householders on behalf of advanced Liberalism.[89] He played a key role in thwarting passage of the Conservative government's 1867 Parks Bill, which aimed to restrict the holding of political meetings and demonstrations in Royal Parks. In 1868 a number of working-class candidates connected with the Reform League received monetary contributions from Mill.[90]

Admittedly, many of the subjects on which Mill spoke in the House of Commons had little to do either with the competition between parties or the distribution of political power. On these, and indeed on some questions that did bear upon party fortunes, the tone and substance of Mill's parliamentary speeches scarcely merit the epithet "extreme." His orations and interventions in the Commons frequently exhibited the deliberation and thoughtfulness befitting a philosopher in politics. When the leader writer in *The Times* complained of Mill's "vehement, narrow partisanship," he mainly had in mind the extra-parliamentary activities undertaken by Mill during his Westminster years. Aside from the reform agitation, the partisanship in question especially concerned the matter and manner of his involvement in the legal and political battles over Jamaican blood and Irish land.

Well known is Mill's prominence in the movement to bring Governor Eyre and certain of his subordinates before the bar of justice for their part in the sanguinary events of October 1865 in Jamaica.[91] Mill believed that the continuance of martial law for several weeks after the suppression of the uprising at Morant Bay had been used by the authorities in the colony not to restore order but to inflict brutal reprisals upon a largely innocent population. A righteous anger gripped Mill when he learned of the summary executions, savage floggings, and wanton destruction of property carried out by agents of the British government. As early as December 1865 he had concluded that nothing before Parliament in 1866, even the Reform Bill itself, could be "more important than the duty of dealing justly with the abominations committed in Jamaica."[92] The findings of the 1866 Royal Commission appointed to investigate the matter were sobering enough to mean that Mill would cleave to his view that Eyre had cruelly unleashed a reign of terror upon the civilian population of Jamaica, for which he must be held accountable.[93]

With the refusal of the Conservative government to bring charges against Eyre and his subordinates, the leadership of the Jamaica Committee—formed to press the government into taking stern measures against those responsible for the abuse of martial law—considered what should be done. Charles Buxton, chairman of the committee, argued that they should aim for no more than an official condemnation of Eyre and financial compensation for the victims or their families. To bring an action of murder against Eyre, Buxton maintained, would antagonize public opinion, which would see Eyre, the upholder of British imperial authority, as the prey of a vindictive persecution. Mill disagreed. Holding that the rule of law had been flagrantly violated, Mill insisted that the committee had no alternative but to launch a prosecution for murder. The fundamental principles of justice and morality demanded nothing less.[94] Buxton resigned the chairmanship and was replaced by Mill.[95] For two years the battle continued, the committee searching in vain for a grand jury prepared to indict Eyre.

Never was Mill more visible in a public cause than he was in his capacity as chairman of the Jamaica Committee, and never was he more uncompromising. For his pains, he excited much hostility. (In the *Autobiography* he refers to the "threats of assassination" he received in the post.)[96] Many felt that his pursuit of Eyre exceeded reasonable bounds. Mill did not waver. In February 1868 he wrote to a correspondent: "This protest & vindication must be made now or never: & to relinquish the effort while a single unexhausted chance remains would be, in my estimation, to make ourselves to some extent participants in the crime."[97]

Mill's staking out of high moral ground on the Jamaica question was entirely consistent with his standing as a liberal public moralist. Readers of his reply to Carlyle's "Occasional Discourse on the Negro Question" (only after Mill's attack did Carlyle retitle it "The Nigger Question"), or of his articles on the Civil War in America, should have expected no less. Yet one is forcibly struck by the depth and intensity of his involvement, properties that owed much to the conspicuous strain of activism he displayed during his parliamentary years. The same can be said of his answer to the Irish land question, as expressed in his 1868 pamphlet *England and Ireland*.

Mill had had plenty of critical things to say, before 1868, about the Irish land system and British rule in Ireland (his *Morning Chronicle* articles of 1846–47 have already been mentioned).[98] Nothing he had said, however, primed the political world for the radicalism of *England and Ireland*, in which Mill for the first time unequivocally advocated fixity of tenure for Irish tenants. Many saw Mill's stark and peremptory plea for a radical solution of the land problem as a call for the virtual expropriation of the Irish landlords.[99] Mill's pamphlet chagrined the propertied and their spokesmen in Parliament and in the press. Considering the substance of Mill's scheme mischievous, they were perhaps equally disturbed by his tone. The pamphlet closed with the following words:

> Let our statesmen be assured that now, when the long deferred day of Fenianism has come, nothing which is not accepted by the Irish tenantry as a permanent solution of the land difficulty, will prevent Fenianism, or something equivalent to it, from being the standing torment of the English Government and people. If without removing this difficulty, we attempt to hold Ireland by force, it will be at the expense of all the character we possess as lovers and maintainers of free government, or respecters of any rights except our own; it will most dangerously aggravate all our chances of misunderstandings with any of the great powers of the world, culminating in war; we shall be in a state of open revolt against the universal conscience of Europe and Christendom, and more and more against our own. And we shall in the end be shamed, or, if not shamed, coerced, into releasing Ireland from the connexion; or we shall avert the necessity only by conceding with the worst grace, and when it will not prevent some generations of ill blood, that which if done at present may still be in time permanently to reconcile the two countries.[100]

In the all but immediate aftermath of the Fenian outbursts, such language from Mill was thought by some to be irresponsible and inflammatory.

Several explanations have been put forward for Mill's startling conversion to fixity of tenure for Irish tenants. E.D. Steele argues that Mill's determination to preserve the Union was behind it. Accepting Fenianism as a genuine but dangerous expression of Irish nationalist aspirations, Mill, in Steele's view, prescribed land reform as the only effective antidote.[101] Lynn Zastoupil stresses Mill's concern for the moral improvement of the Irish people, and suggests that he had come to perceive both Irish nationalism and fixity of tenure as agencies of moral development.[102] I have said elsewhere that Mill, impelled by Fenianism to tackle the issue, seized upon the Irish land question as a decisive test of English moral will: the elevation of the Irish masses had in his mind become fused to the moral condition of England.[103] Whatever one makes of these respective interpretations, there can be no doubt that the fuel powering *England and Ireland* carries a high concentration of the intense political activism that defined this phase of Mill's life.

In the *Autobiography* Mill abjures the notion, allegations notwithstanding, that "in the trials of public life" he had shown himself to be "intemperate and passionate."[104] (It is worth noting that he did not claim to have been temperate and dispassionate.) Yet he had no reason to wonder at the reaction he had provoked. It is perfectly understandable that people who had not felt threatened by Mill the theoretician might have had misgivings about Mill the political activist. Had they been in a position to survey Mill's experience of political engagement over a period of decades, they might have been less surprised by the conduct he evinced during his Westminster years. The philosophical vocation, as understood by Mill, enjoined circumspect investigation and analysis. The method, for which he had an obvious aptitude, could impose constraints on the conclusions reached, though it did not invariably do so. The result might cast into the shade a radical moral consciousness that nonetheless left its mark on much that Mill wrote. When aroused, by whatever stimuli, his radical political consciousness—a subset of that moral consciousness—expressed itself practically in modes of action that were largely exempt from such constraints. Theory generally blunted Mill's radical edge; practice sharpened it.

NOTES

Introduction

1. Michael St. John Packe, *The Life of John Stuart Mill* (New York: Macmillan, 1954); Nicholas Capaldi, *John Stuart Mill: A Biography* (Cambridge: Cambridge University Press, 2004).
2. *The Collected Works of John Stuart Mill* (henceforth cited as *CW*), vol. 1, *Autobiography and Literary Essays*, ed. John M. Robson and Jack Stillinger (Toronto: University of Toronto Press, 1981), Appendix G, 612–13.
3. Bruce Mazlish, *James and John Stuart Mill: Father and Son in the Nineteenth Century* (New York: Basic Books, 1975); P.J. Glassman, *J.S. Mill: The Evolution of a Genius* (Gainesville: University of Florida Press, 1985).
4. *CW*, vol. 1, *Autobiography and Literary Essays*, 79.
5. Janice Carlisle, *John Stuart Mill and the Writing of Character* (Athens, Georgia: University of Georgia Press, 1991), 99.
6. F.A. Hayek, *John Stuart Mill and Harriet Taylor: Their Friendship and Subsequent Marriage* (London: Routledge & Kegan Paul, 1951), chapters 2 and 3.
7. Packe, *Life of John Stuart Mill*, 115–32, 137–54.
8. Capaldi, *John Stuart Mill*, 102–11, 113–17.
9. Josephine Kamm, *John Stuart Mill in Love* (London: Gordon and Cremonesi, 1977).
10. Phyllis Rose, *Parallel Lives: Five Victorian Marriages* (New York: Knopf, 1984), 95–140.
11. Jo Ellen Jacobs, *The Voice of Harriet Taylor Mill* (Bloomington, Indiana: Indiana University Press, 2002).
12. Some significant contributions to this literature—in addition to the works by Hayek, Packe, Rose, and Jacobs cited above—include: H.O. Pappé, *John Stuart Mill and the Harriet Taylor Myth* (Melbourne: Melbourne University Press, 1960); J.M. Robson, "Harriet Taylor and John Stuart Mill: Artist and Scientist," *Queen's Quarterly* 73 (1966): 167–86; Jack Stillinger, "Who Wrote J.S. Mill's Autobiography?" *Victorian Studies* 27 (1983): 7–23.
13. See, for example, Julia Annas, "Mill and *The Subjection of Women*," *Philosophy* 52 (1977): 179–94; Barbara Caine, "John Stuart Mill and the English Women's Movement," *Historical Studies* 18 (1978): 52–67; Mary L. Shanley, "Marital Slavery and Friendship: John Stuart Mill's *The Subjection of Women*," *Political Theory* 9 (1981): 229–47; Gail Tulloch, *Mill and Sexual Equality* (Hemel Hempstead: Harvester Wheatleaf, 1989); Susan Mendus, "John Stuart Mill and Harriet Taylor on Women and Marriage," *Utilitas* 6 (1994): 287–99; Maria H. Morales, *Perfect Equality: John Stuart Mill on Well-Constituted Communities* (Lanham, Maryland: Rowman & Littlefield, 1996).
14. Fred Kaplan, *Thomas Carlyle: A Biography* (Ithaca, New York: Cornell University Press, 1983).
15. Emery Neff, *Carlyle and Mill: Mystic and Utilitarian* (New York: Columbia University Press, 1924); republished as *Carlyle and Mill: An Introduction to Victorian Thought* (New York: Columbia University Press, 1926).

Notes

16. R.K. Pankhurst, *The Saint Simonians, Mill and Carlyle: A Preface to Modern Thought* (London: Sidgwick and Jackson, 1957).
17. John Skorupski, ed., *The Cambridge Companion to Mill* (Cambridge: Cambridge University Press, 1998).

One   The Father, the Son, and the Manly Spirit

1. *CW*, vol. 1, *Autobiography and Literary Essays*, 5.
2. Jo Ellen Jacobs, ed., *The Complete Works of Harriet Taylor Mill* (Bloomington, Indiana; Indiana University Press, 1998), 375.
3. *CW*, vol. 1, *Autobiography and Literary Essays*, 175–77. Mill's formal treatment of the problem is found in *CW*, vols. 7–8, *A System of Logic Ratiocinative and Inductive*, ed. John M. Robson (Toronto: University of Toronto Press, 1973), vol. 8, 836–43.
4. The literature on the difficulties presented by Mill's *Autobiography* is substantial. The field is headed by Jack Stillinger, whose relevant publications include: "The Text of John Stuart Mill's *Autobiography*," *Bulletin of the John Rylands Library* 43 (1960): 220–42; his edition of *The Early Draft of John Stuart Mill's Autobiography* (Urbana, Illinois: University of Illinois Press, 1961); "Who Wrote J.S. Mill's Autobiography," *Victorian Studies* 27 (1983), 7–23; and "John Mill's Education: Fact, Fiction and Myth," in *A Cultivated Mind: Essays on J.S. Mill Presented to John M. Robson*, ed. Michael Laine (Toronto: University of Toronto Press, 1991), 19–43. A number of other scholars have made significant contributions to the discussion: Robert D. Cumming, "Mill's History of His Ideas," *Journal of the History of Ideas* 25 (1964): 235–56; William Thomas, "John Stuart Mill and the Uses of Autobiography," *History* 56 (1971): 341–59; Alan Ryan, *J.S. Mill* (London: Routledge & Kegan Paul, 1974), chapter 1; James McDonnell, "Success and Failure: A Rhetorical Study of the First Two Chapters of John Stuart Mill's *Autobiography*," *University of Toronto Quarterly* 45 (1976): 109–22; Jonathan Loesberg, *Fictions of Consciousness: Mill, Newman, and the Reading of Victorian Prose* (New Brunswick, New Jersey: Rutgers University Press, 1986); John M. Robson, introduction to the Penguin edition of Mill's *Autobiography* (London: Penguin, 1989). There is also much of interest on the *Autobiography* in Janice Carlisle's *John Stuart Mill and the Writing of Character*, and in Regenia Gagnier, *Subjectivities: A History of Self-Representation in Britain, 1832–1920* (Oxford: Oxford University Press, 1991).
5. The most ambitious study of the father/son relationship is Mazlish's *James and John Stuart Mill*; its psychoanalytical interpretation has generally not found favor among Mill scholars.
6. Alexander Bain, *James Mill: A Biography* (London: Longmans, Green, 1882), 3–5.
7. Graham Wallas, *The Life of Francis Place, 1771–1854* (London: Longmans, Green, 1898), 70n–71n.
8. *CW*, vol. 1, *Autobiography and Literary Essays*, 5.
9. Bain, *James Mill*, 59.
10. Bain, *James Mill*, 60. Francis Place, who came to know the Mills in 1817, immediately recognized the incompatibility of the couple: "Mrs. Mill is a patient, quiet soul, hating wrangling, and although by no means meanly submissive, manages to avoid quarrelling in a very admirable manner." Wallas, *Life of Francis Place*, 73.
11. See James Mill's essay "Education" (1819), in *Essays* (London: printed Innes [1825]), and his *Analysis of the Phenomena of the Human Mind*, 2 vols. (London: Baldwin and Cradock, 1829).
12. Anna J. Mill, "The Education of John—Some Further Evidence," *Mill News Letter* 11 (Winter, 1976): 11. Having stood no chance of winning the race for Wilhelmina's hand, he welcomed the prospect of competing with William Forbes on a level playing field when it came to "the education of a son."

13. Anna J. Mill, "The Education of John," 11–12. This letter, dated 26 February 1820, was written to a Colonel Walker; for a valuable record of the readings that made up this education, see *CW*, vol. 1, *Autobiography and Literary Essays*, Appendix B, 551–81.
14. *CW*, vol. 1, *Autobiography and Literary Essays*, 33.
15. *CW*, vol. 1, *Autobiography and Literary Essays*, 35.
16. Wallas, *Life of Francis Place*, 74.
17. See Bain, *James Mill*, 95–96.
18. Elie Halévy, *The Growth of Philosophic Radicalism*, trans. Mary Morris (London: Faber and Gwyer, 1928), 251.
19. In a letter of July 1812, written when John Mill was six years old, Bentham, who had himself been a child prodigy, offered to become the boy's guardian and complete his education in the event of James Mill's dying during John's childhood. James Mill gratefully accepted this offer, stipulating that its implementation, if necessary, "shall be as good as possible; and then we may perhaps leave a successor worthy of both of us." Bain, *James Mill*, 119–20.
20. The analysis and categorization of motive claimed Bentham's attention in both *An Introduction to the Principles of Morals and Legislation* (1780) and *A Table of the Springs of Action* (1817).
21. Quoted in John Dinwiddy, *Bentham* (Oxford: Oxford University Press, 1989), 24.
22. See Halévy, *Growth of Philosophic Radicalism*, 403–12.
23. Dinwiddy, *Bentham*, 30.
24. See John Dinwiddy, "Bentham's Transition to Political Radicalism, 1809–10," *Journal of the History of Ideas* 35 (1975): 683–700.
25. *CW*, vol. 1, *Autobiography and Literary Essays*, 67.
26. *CW*, vol. 1, *Autobiography and Literary Essays*, 67, 69.
27. "If I had been by nature extremely quick of apprehension, or had possessed a very accurate and retentive memory, or were of remarkably active and energetic character, the trial would not be conclusive; but in all these natural gifts I am rather below than above par." *CW*, vol. 1, *Autobiography and Literary Essays*, 33.
28. *CW*, vol. 1, *Autobiography and Literary Essays*, 31.
29. Anna J. Mill, "The Education of John," 11–12.
30. *CW*, vol. 1, *Autobiography and Literary Essays*, 33.
31. Quoted in William Thomas, *The Philosophic Radicals: Nine Studies in Theory and Practice, 1817–1841* (Oxford: Clarendon Press, 1979), 132.
32. Anna J. Mill, "The Education of John," 13.
33. *CW*, vol. 1, *Autobiography and Literary Essays*, 49.
34. *CW*, vol. 1, *Autobiography and Literary Essays*, 49.
35. The case for the continuing influence of this upbringing on James Mill's moral outlook has been powerfully set forth by William Thomas in his chapter on James Mill in *Philosophic Radicals*, 95–146, especially at 98–100. For a valuable investigation of James Mill's "civil religion," see Terence Ball, *Reappraising Political Theory* (Oxford: Oxford University Press, 1995), 131–57.
36. For a comprehensive discussion of Mill's response to Socrates and ancient Athens, see Nadia Urbinati, *Mill on Democracy: From the Athenian Polis to Representative Government* (Chicago: University of Chicago Press, 2002). See also F.E. Sparshott, introduction to *CW*, vol. 11, *Essays on Philosophy and the Classics*, ed. John M. Robson (Toronto: University of Toronto Press, 1978); Frank Turner, *The Greek Heritage in Victorian Britain* (New Haven: Yale University Press, 1981); Geraint Williams, "J.S. Mill on the Greeks: History Put to Use," *Mill News Letter* 17 (Winter 1982): 1–11; Eugenio Biagini, "John Stuart Mill and the Model of Ancient Athens," in *Citizenship and Community: Liberals, Radicals, and Collective Identities in the British Isles, 1865–1931*, ed. Eugenio Biagini (Cambridge: Cambridge University Press, 1996), 21–44; and T.H. Irwin, "Mill and the Classical World," in *Cambridge Companion to Mill*, ed. Skorupski, 423–63.
37. *CW*, vol. 1, *Autobiography and Literary Essays*, 41.

38. *CW*, vol. 1, *Autobiography and Literary Essays*, 45.
39. *CW*, vol. 1, *Autobiography and Literary Essays*, 45, 47.
40. *CW*, vol. 1, *Autobiography and Literary Essays*, 47.
41. *CW*, vols. 28–29, *Public and Parliamentary Speeches*, ed. John M. Robson and Bruce L. Kinzer (Toronto: University of Toronto Press, 1988), vol. 28, 38.
42. *CW*, vols. 14–17, *The Later Letters of John Stuart Mill, 1849–1873*, ed. Francis E. Mineka and Dwight N. Lindley (Toronto: University of Toronto Press, 1972), vol. 16, 1483; *The Times*, 11 November 1868, 5.
43. For Mill's *Three Essays on Religion*, see *CW*, vol. 10, *Essays on Ethics, Religion and Society*, ed. John M. Robson (Toronto: University of Toronto Press, 1969), 369–489.
44. Samuel Bentham knew something of John Mill's accomplishments before his arrival in France, chiefly from what Jeremy said in his letters to his brother. When the plan for sending John over was taking on a serious look, Jeremy told Samuel that when John turned up Samuel could "shew him for 6d a piece and get rich." *The Correspondence of Jeremy Bentham*, vol. 9, ed. Stephen Conway (Oxford: Clarendon Press, 1989), 380–81. James and John Mill had in fact met Sir Samuel on two occasions before John's stay in France. In 1813 they had visited him at his house in Gosport when they were doing a tour of southwest England; the following year they saw the entire family at Ford Abbey. See *CW*, vol. 1, *Autobiography and Literary Essays*, 55, 57. In later life George Bentham recalled the experience of meeting John Mill in August 1813, when the latter was seven. "He was then in some respects a prodigy. He had a wonderfully precocious mind; his father, a cold Scotchman with more ability than principle, whilst neglecting his wife and younger children, took the greatest pains in developing John's mind without caring for his manners. At this time at the age of six [sic] he was a Greek and Latin scholar, a historian, and a logician, and fond of showing off his proficiency without the slightest reserve." Marion Filipiuk, ed., *George Bentham: Autobiography, 1800–1834* (Toronto: University of Toronto Press, 1997), 9.
45. *CW*, vol. 1, *Autobiography and Literary Essays*, 37.
46. *CW*, vols. 12–13, *The Earlier Letters of John Stuart Mill, 1812–1848*, ed. Francis E. Mineka (Toronto: University of Toronto Press, 1963), vol.13, 540. George Bentham, in his autobiography, indicates that John Mill's hosts were pleased with their guest: "he struck us much by the quickness of his perceptions and the powers of his mind . . . I find, among my notes made at the time memoranda of his rapid progress in French, of his readiness at difficult algebraic problems . . . His visit was also on many counts a pleasant one to us, and we had every reason to be satisfied with his conduct, disposition and principles during the seven or eight months he remained with us." Filipiuk, ed., *George Bentham: Autobiography*, 63.
47. See Anna J. Mill, ed., *John Mill's Boyhood Visit to France: A Journal and Notebook* (Toronto: University of Toronto Press, 1960). The editor provides a valuable introduction to this journal.
48. *CW*, vol. 1, *Autobiography and Literary Essays*, Appendix G, 612–13.
49. *CW*, vols. 26–27, *Journals and Debating Speeches*, ed. John M. Robson (Toronto: University of Toronto Press, 1988), vol. 27, 642. When J.S. Mill edited and reissued, in 1869, his father's *Analysis of the Phenomena of the Human Mind*, he publicly voiced the sentiments he had confided to his journal fifteen years before. His preface to this edition included the following passage: "It is to the author of the present volumes that the honour belongs of being the reviver and second founder of the Association psychology. Great as is this merit, it was but one among many services which he rendered to his generation and mankind. When the literary and philosophical history of this century comes to be written as it deserves to be, very few are the names figuring in it to whom as high a place will be awarded as to James Mill. In the vigour and penetration of his intellect he has had few superiors in the history of thought: in the wide compass of the human interests which he cared for and served, he was almost equally remarkable: and the energy and determination of his character, giving effect to as single-minded an ardour for the improvement of mankind and of human life as I believe has ever existed, make his life a memorable example." *CW*, vol. 31, *Miscellaneous Writings*, ed. John M. Robson

NOTES

    (Toronto: University of Toronto Press, 1989), 99. No account of "the literary and philosophical history" of the nineteenth century endorses J.S. Mill's view of his father's importance.
50. *CW*, vol. 1, *Autobiography and Literary Essays*, 53.
51. *The Works and Correspondence of David Ricardo*, ed. Piero Sraffa, vol. 7, *Letters 1816–1818* (Cambridge: Cambridge University Press, 1952), 301–02.
52. *CW*, vol. 1, *Autobiography and Literary Essays*, 53.
53. *CW*, vol. 1, *Autobiography and Literary Essays*, 53.
54. *CW*, vol. 1, *Autobiography and Literary Essays*, 55.
55. *CW*, vol. 1, *Autobiography and Literary Essays*, 33.
56. *CW*, vol. 1, *Autobiography and Literary Essays*, 31.
57. *CW*, vol. 1, *Autobiography and Literary Essays*, 65.
58. *CW*, vol. 1, *Autobiography and Literary Essays*, 67.
59. *CW*, vol. 1, *Autobiography and Literary Essays*, 77, 79. For valuable studies of Austin, see W.L. Morison, *John Austin* (Stanford, California: Stanford University Press, 1982), and Lotte and Joseph Hamburger, *Troubled Lives: John and Sarah Austin* (Toronto: University of Toronto Press, 1985).
60. Sarah Austin's tendency to gossip about Mill's "friendship" with Harriet Taylor subsequently soured his relations with Mrs. Austin. For Mill's disparaging comments in the "Early Draft," see *CW*, vol. 1, *Autobiography and Literary Essays*, 186.
61. John Clive observes that "Austin's strong personality appears to have been one of the few by which even Macaulay, always formidable himself, was awed." See John Clive, *Macaulay: The Shaping of the Historian* (New York: Knopf, 1974), 63.
62. *CW*, vol. 1, *Autobiography and Literary Essays*, 79.
63. *CW*, vol. 1, *Autobiography and Literary Essays*, 81.
64. See Percy Craddock, *Recollections of the Cambridge Union 1815–1939* (Cambridge: Cambridge University Press, 1953), 169–70.
65. *CW*, vol. 1, *Autobiography and Literary Essays*, 79.
66. *CW*, vol. 1, *Autobiography and Literary Essays*, 67.
67. *CW*, vol. 1, *Autobiography and Literary Essays*, 69.
68. *CW*, vol. 1, *Autobiography and Literary Essays*, 71.
69. James Mill, *Analysis of the Phenomena of the Human Mind*, 2 vols. (London: Baldwin and Cradock, 1829), vol. 1, 70.
70. *CW*, vol. 1, *Autobiography and Literary Essays*, 81; Mill persuaded Bentham's then amanuensis, Richard Doane, to join the group, and Doane secured a room in Bentham's house for the purpose.
71. Mill gave a playful and detailed account of recent happenings "in the Utilitarian world" in a letter to George and Harriet Grote dated 1 September 1824. See *CW*, vol. 32, *Additional Letters of John Stuart Mill*, ed. Marion Filipiuk, Michael Laine, and John M. Robson (Toronto: University of Toronto Press, 1991), 1–7.
72. This capacity for acting on his beliefs evidently extended to the dissemination of birth control pamphlets that got him into a scrape in 1823. For discussions of the incident, and its long-lived repercussions, see W.D. Christie, *John Stuart Mill and Abraham Hayward, QC: A Reply about Mill to a Letter to the Rev. Stopford Brooke, Privately Circulated and Actually Published* (London: King, 1873); N.E. Himes, "John Stuart Mill's Attitude Toward Neo-Malthusianism," Suppl. to the *Economic Journal* 4 (1929): 457–84; Pedro Schwartz, *The New Political Economy of J.S. Mill* (London: Weidenfeld and Nicolson, 1972), Appendix 2, 245–56; F.E. Mineka, "John Stuart Mill and Neo-Malthusianism, 1873," *Mill News Letter* 8 (Fall, 1972), 3–10.
73. *CW*, vol. 1, *Autobiography and Literary Essays*, 197; of course this is said in the section where he seeks to contrast his own limitations with Harriet Taylor's prodigious range of abilities.
74. *CW*, vol. 1, *Autobiography and Literary Essays*, 65, 67.
75. *CW*, vol. 12, *Earlier Letters*, 9n.
76. From 1858 until his death Mill spent six months of each year in Avignon, where Harriet was buried. For twelve of these years, Napoleon III, whose regime Mill loathed, was in power. He

did not interpret this regrettable dominance to mean, however, that the French people had become "quiet and contented slaves."
77. *CW*, vol. 1, *Autobiography and Literary Essays*, 336.
78. *CW*, vol. 1, *Autobiography and Literary Essays*, 83. The accuracy of this observation is confirmed by Company records—see Martin Moir's introduction to *CW*, vol. 30, *Writings on India*, ed. John M. Robson, Martin Moir, and Zawahir Moir (Toronto: University of Toronto Press, 1990), xiii.
79. Bain, *James Mill*, 207.
80. *CW*, vol. 1, *Autobiography and Literary Essays*, 85.
81. *CW*, vol. 30, *Writings on India*, xvii–xviii.
82. *CW*, vol. 1, *Autobiography and Literary Essays*, 85.
83. *CW*, vol. 1, *Autobiography and Literary Essays*, 84n.
84. *CW*, vol. 1, *Autobiography and Literary Essays*, 87. A number of able scholars ascribe far greater significance to Mill's India House career than he himself did. The standard work in this sphere: Lynn Zastoupil, *John Stuart Mill and India* (Stanford, California: Stanford University Press, 1994); see also the essay collection, *J.S. Mill's Encounter with India*, ed. Martin I. Moir, Douglas M. Peers, and Lynn Zastoupil (Toronto: University of Toronto Press, 1999). For a valuable treatment of the problem that more closely fits Mill's own assessment, see Trevor Lloyd, "John Stuart Mill and the East India Company," in *A Cultivated Mind*, ed. Laine, 44–79.
85. According to Janice Carlisle, Mill was so far from "indifferent" to this exclusion that his consciousness of it greatly contributed to his crisis of the second half of the 1820s. Although I do not find her interpretation persuasive, she unquestionably argues her case with notable force. See Carlisle, *John Stuart Mill and the Writing of Character*, 63–75.
86. *CW*, vol. 1, *Autobiography and Literary Essays*, 93.
87. *CW*, vol. 1, *Autobiography and Literary Essays*, 85.
88. For an excellent examination of Mill's early journalism, see Ann P. Robson's introduction to *CW*, vols. 22–25, *Newspaper Writings*, ed. Ann P. Robson and John M. Robson (Toronto: University of Toronto Press, 1986), vol. 22, xxxi–xli; for Mill's debating speeches, see the penetrating analysis of John M. Robson in his introduction to *CW*, vols. 26–27, *Journals and Debating Speeches*, vol. 26, xviii–xliii.
89. *CW*, vol. 22, *Newspaper Writings*, 44.
90. Roebuck noted in his autobiographical account: "I became . . . a pupil of John Mill, who, although younger than myself, was far in advance of me in philosophy and politics." R.E. Leader, ed., *Life and Letters of John Arthur Roebuck, with Chapters of Autobiography* (London: Arnold, 1897), 28.
91. Speaking of his friendship with Mill and Graham, Roebuck observed that his "intimacy [with Mill] increased day by day, and was strengthened by the fact that Graham and myself became sworn friends—brothers, in fact—and with John Mill formed a triumvirate which we laughingly called the 'Trijackia,' all of us being named John." Leader, ed., *Life and Letters of John Arthur Roebuck*, 28.
92. Alexander Bain, *John Stuart Mill, a Criticism: With Personal Recollections* (London: Longmans, Green, 1882), 39–40.
93. The record of this episode derives from "some memoranda of conversations" Bain "had with Roebuck not long before his death"; from "a recollection" Bain elicited from J.S. Mill's surviving siblings; and from Roebuck's published reminiscences, which states that James Mill "took occasion to remark to myself especially, that he had no great liking for his son's new friends. I, on the other hand, let him know that I had no fear of him who was looked upon as a sort of Jupiter Tonans." Roebuck remembered that John Mill protested against his father's treatment of his friends, "but the result was that we soon ceased to see John Mill at his home." Leader, ed., *Life and Letters of John Arthur Roebuck*, 28–29.
94. *CW*, vol. 1, *Autobiography and Literary Essays*, 143.
95. *CW*, vol. 1, *Autobiography and Literary Essays*, 137, 139.

96. Only those who know something of Bentham's method and handwriting can begin to appreciate what this project entailed. Mill's discussion of the matter in the *Autobiography*, however, gives some idea of the magnitude of the undertaking. "Mr. Bentham had begun this treatise three times, at considerable intervals, each time in a different manner, and each time without reference to the preceding: two of the three times he had gone over nearly the whole subject. These three masses of manuscript it was my business to condense into a single treatise; adopting the one last written as the groundwork, and incorporating with it as much of the other two as it had not completely superseded. I had also to unroll such of Bentham's involved and parenthetical sentences, as seemed to overpass by their complexity the measure of what readers were likely to take the pains to understand. It was further Mr. Bentham's particular desire that I should, from myself, endeavour to supply any *lacunæ* which he had left; and at his instance I read, for this purpose, the most authoritative treatises on the English Law of Evidence, and commented on a few of the objectionable points of the English rules, which had escaped Bentham's notice. I also replied to the objections which had been made to some of his doctrines, by reviewers of Dumont's book, and added a few supplementary remarks on some of the more abstract parts of the subject, such as the theory of improbability and impossibility." *CW*, vol. 1, *Autobiography and Literary Essays*, 117.
97. From Martin Moir's introduction to *CW*, vol. 30, *Writings on India*, xv.
98. *CW*, vol. 1, *Autobiography and Literary Essays*, 121, 123.
99. *CW*, vol. 1, *Autobiography and Literary Essays*, 141.
100. *CW*, vol. 1, *Autobiography and Literary Essays*, 141.
101. *CW*, vol. 1, *Autobiography and Literary Essays*, 145.
102. See, especially, A.W. Levi, "The 'Mental Crisis' of John Stuart Mill," *Psychoanalytic Review* 32 (1945): 86–101, and Mazlish, *James and John Stuart Mill*.
103. *CW*, vol. 1, *Autobiography and Literary Essays*, 147.
104. *CW*, vol. 1, *Autobiography and Literary Essays*, 151.
105. *CW*, vol. 1, *Autobiography and Literary Essays*, 147.
106. *CW*, vol. 1, *Autobiography and Literary Essays*, 13.
107. For his reference to music, see *CW*, vol. 1, *Autobiography and Literary Essays*, 147.
108. *CW*, vol. 1, *Autobiography and Literary Essays*, 151.
109. *CW*, vol. 1, *Autobiography and Literary Essays*, 147.
110. *CW*, vol. 1, *Autobiography and Literary Essays*, 145, 147.
111. *CW*, vol. 1, *Autobiography and Literary Essays*, 49.
112. *CW*, vol. 1, *Autobiography and Literary Essays*, 145.
113. *CW*, vol. 1, *Autobiography and Literary Essays*, 49.

Two   Gathering Truths, 1826–30

1. See Norman Gash, *Aristocracy and People: Britain 1815–1865* (Cambridge, Massachusetts: Harvard University Press), 21–22.
2. In 1816 Parliament declined to renew the income tax, the wars against Napoleon having ended the previous year. As for the matter of status and income, Boyd Hilton notes that "in order to have pretensions to *upper*-middle-class gentility it was necessary to have an income of at least £200 per annum and preferably £300." See Boyd Hilton, *A Mad, Bad, & Dangerous People? England 1783–1846* (Oxford: Oxford University Press, 2006), 126, 128.
3. *The Collected Letters of Thomas and Jane Welsh Carlyle*, vol. 9, ed. Charles Richard Sanders (Durham, North Carolina: Duke University Press, 1981), 20.
4. Quoted in Gash, *Aristocracy and People*, 24.
5. *CW*, vols. 18–19, *Essays on Politics and Society*, ed. John M. Robson (Toronto: University of Toronto Press, 1977), vol. 18, 193.

6. *The Collected Letters of Thomas and Jane Welsh Carlyle*, vol. 5, ed. Charles Richard Sanders (Durham, North Carolina: Duke University Press, 1976), 398.
7. Caroline Fox, *Memories of Old Friends, Being Extracts from the Journals and Letters of Caroline Fox, from 1835 to 1871*, ed. H.N. Pym, 2nd ed., 2 vols. (London: Smith, Elder, 1882), vol. 1, 138.
8. Henry Taylor, *Autobiography*, 2 vols. (London: Longmans, Green, 1885), vol. 1, 78.
9. *CW*, vol. 1, *Autobiography and Literary Essays*, 39.
10. Quoted in Hayek, *John Stuart Mill and Harriet Taylor*, 32.
11. Henry Solly, *These Eighty Years*, 2 vols. (London: Simpkin and Marshall, 1893), vol. 1, 147.
12. See Anna J. Mill, "John Stuart Mill's Visit to Wordsworth, 1831," *Modern Language Review* 44 (1949): 342.
13. Henry Taylor, *Autobiography*, vol. 1, 159.
14. Charles Greville, *The Greville Memoirs*, 8 vols., ed. Lytton Strachey and Roger Fulford (London: Macmillan, 1938), vol. 2, 58.
15. Bain, *John Stuart Mill*, 189.
16. *Collected Letters of Thomas and Jane Welsh Carlyle*, vol. 5, 398.
17. Henry Taylor, *Correspondence of Henry Taylor*, ed. Edward Dowden (London: Longmans, Green, 1888), 28. In his *Autobiography* Taylor both acknowledges Mill's conversational power and notes an intense deliberateness on the part of the speaker—a characteristic that no doubt affected Greville's estimate. Taylor said that Mill "took his share in conversation, and talked ably and well of course but with such a scrupulous solicitude to think exactly what he should say and say exactly what he thought, that he spoke with an appearance of effort and as if with an impediment of mind." Henry Taylor, *Autobiography*, vol. 1, 79.
18. Mill's approach to conversation had nothing in common with the great talkers of his age—Sydney Smith, Carlyle, and Macaulay, for example—who preferred monologue to dialogue. Smith, when asked about the caliber of Macaulay's performance after the two of them had entered the conversational ring together, said his adversary had been "unusually brilliant, some splendid flashes of silence." Mill, Bain tells us, enjoyed this story, and offered one of his own about "two Frenchmen of this species, pitted against each other. One was in full possession, but so intent was the other upon breaking in, that a third person watching the contest, exclaimed, 'If he spits, he's done.'" Bain, *John Stuart Mill*, 189n. Bain avers (188) that Mill "had humour and lightness, and did not restrain their display."
19. Henry Cole's Diary is located in the National Art Library, the Victoria and Albert Museum.
20. Cole's Diary refers frequently to such visits to India House during these years. Cole went on to become one of the most distinguished civil servants of his generation. His career included central involvement in the mounting of the Great Exhibition of 1851 and the founding of what became the Victoria and Albert Museum.
21. Cole's entry for 4 September 1828 notes: "Drank tea with John Mill & employed the evening in the examination of his Botanical Specimens of which his liberality made me several presents Sundries, etc." The entry for a fortnight later similarly reads: "Passed the evening and drank tea with John Mill who most liberally from his collection of Plants contributed to mine."
22. Cole's Diary, 19 November 1828.
23. Cole's Diary, 29 June 1830. In January 1831 Mill sent to William Jackson Hooker, the director of Kew Gardens, a specimen of "Lilium martagon, a plant new to the British flora, but certainly wild, & as far as it is possible to judge, indigenous." *CW*, vol. 12, *Earlier Letters*, 69.
24. *CW*, vol. 1, *Autobiography and Literary Essays*, 85n.
25. Cole's Diary, 19 April 1829.
26. Cole's Diary, 3 February 1830.
27. There are five extant journals of John Mill's walking tours. Apart from the two already mentioned, these pertain to Sussex for some ten days in the summer of 1827; to Berkshire, Buckinghamshire, Oxfordshire, and Surrey in the first half of July 1828; and to western Cornwall for a week in October 1832. See *CW*, vol. 27, *Journals and Debating Speeches*, 455–637.

28. Lotte and Joseph Hamburger, *Troubled Lives*, 65.
29. *CW*, vol. 1, *Autobiography and Literary Essays*, 83. In 1830–31 Mill and Graham had the idea of collaborating on a series of essays dealing with various technical aspects of economic theory; this became a solo venture for Mill when Graham took issue with certain lines of argument being advanced by his friend. See *CW*, vol. 1, *Autobiography and Literary Essays*, 125.
30. *CW*, vol. 1, *Autobiography and Literary Essays*, 83, 99, 105, 161.
31. *CW*, vol. 32, *Additional Letters*, 3.
32. *CW*, vol. 1, *Autobiography and Literary Essays*, 158.
33. *CW*, vol. 1, *Autobiography and Literary Essays*, 161.
34. *CW*, vol. 1, *Autobiography and Literary Essays*, 133.
35. *The Collected Works of Samuel Taylor Coleridge*, vol. 4, ed. Kathleen Coburn (London and Princeton, New Jersey: Princeton University Press, 1969), 500.
36. See Merrill Distad, *Guessing at Truth: The Life of Julius Charles Hare* (Shepherdstown, West Virginia: Patmos, 1979), 39.
37. Distad, *Guessing at Truth*, 47.
38. See John M. Robson's introduction to *CW*, vols. 26–27, *Journals and Debating Speeches*, vol. 26, xxxi–xxxii; for Mill's speech, see *Journals and Debating Speeches*, vol. 26, 434–42. Richard Monckton Milnes, another Cambridge man and Apostle, was present for the debate, and wrote to his father that "Sterling spoke splendidly, and Mill made an essay on Wordsworth's poetry for two and three-quarter hours, which delighted me." Thomas Wemyss Reid, *The Life, Letters, and Friendships of Richard Monckton Milnes, First Lord Houghton*, 2 vols. (London: Cassell, 1890), vol. 1, 62.
39. *CW*, vol. 26, *Journals and Debating Speeches*, 441.
40. *CW*, vol. 1, *Autobiography and Literary Essays*, 163.
41. Quoted in Kaplan, *Thomas Carlyle*, 373.
42. *CW*, vol. 1, *Autobiography and Literary Essays*, 161, 163.
43. Circumstances did not allow Mill and Sterling to see much of each other. By 1830 Sterling had become involved in a high-minded yet crackpot scheme aimed at overthrowing the tyrannical regime of Spain's Ferdinand VII. The ill-fated invasion to which it gave rise resulted in the summary execution of many of Sterling's coconspirators. Sterling himself held back, his romantic impulses having been diverted in a non-Iberian direction shortly before the expedition sailed. He had fallen deeply in love with the sister of a fellow Apostle, and he married her in early November 1830. Shortly thereafter he suffered a serious pulmonary attack, and this led him to accept an invitation to manage a West Indian sugar estate in which he and his family had an interest. Although he returned to England in August 1832, he spent little time in London in the several years that followed. None of this kept Mill and Sterling from sharing thoughts and feelings with one another in the letters they exchanged.
44. *CW*, vol. 10, *Essays on Ethics, Religion and Society*, 132.
45. *CW*, vol. 10, *Essays on Ethics, Religion and Society*, 138. This quotation adheres to the 1840 text of Mill's "Coleridge"; see variants at foot of 138.
46. *CW*, vol. 12, *Earlier Letters*, 42.
47. *CW*, vol. 1, *Autobiography and Literary Essays*, 161.
48. Richard Holmes, *Coleridge: Darker Reflections, 1804–1834* (New York: Pantheon, 1998), 281n.
49. *CW*, vol. 1, *Autobiography and Literary Essays*, 123.
50. Kant is not mentioned in Mill's *Autobiography*, and of the four hundred plus pages of the "Index of Persons" compiled for Mill's entire corpus, Kant and his writings claim less than half a page; see *CW*, vol. 33, *Indexes to the Collected Works of John Stuart Mill*, ed. Jean O'Grady (Toronto: University of Toronto Press, 1991), 229–30.
51. With regard to Carlyle's articles on German literature written for the *Edinburgh Review* and the *Foreign Review and Continental Miscellany*, Mill stated that "for a long time I saw nothing in them . . . but insane rhapsody." *CW*, vol. 1, *Autobiography and Literary Essays*, 169.
52. *CW*, vol. 1, *Autobiography and Literary Essays*, 169.

53. *CW*, vol. 12, *Earlier Letters*, 84.
54. See Mill's speech on "Perfectibility," in *CW*, vol. 26, *Journals and Debating Speeches*, 429–30. Mill stated: "I will even say, that so far from its being a mark of wisdom to despair of human improvement there is no more certain indication of narrow views and a limited understanding, and that the wisest men of all political and religious opinions, from Condorcet to Mr. Coleridge, have been something nearly approaching to perfectibilians."
55. *Collected Works of Samuel Taylor Coleridge*, vol. 10, *On the Constitution of the Church and the State*, ed. John Colmer (London and Princeton, New Jersey: Princeton University Press, 1976), xiii.
56. *Collected Works of Samuel Taylor Coleridge*, vol. 6, *Lay Sermons*, ed. R.J. White (London and Princeton, New Jersey: Princeton University Press, 1972), 216–17.
57. *Collected Works of Samuel Taylor Coleridge*, vol. 6, *Lay Sermons*, 221–23.
58. *Collected Works of Samuel Taylor Coleridge*, vol. 10, *On the Constitution of Church and State*, 42–43.
59. *CW*, vol. 12, *Earlier Letters*, 75–76.
60. Joseph Hume, Francis Place, George Grote, and Henry Brougham, among others, also played an important part in the creation of the University of London.
61. Lotte and Joseph Hamburger, *Troubled Lives*, 33.
62. *CW*, vol. 1, *Autobiography and Literary Essays*, 185.
63. In a letter to Gustave d'Eichtal of May 1829, Mill alluded to "the very worst point in our national character, the disposition to sacrifice every thing to accumulation, & that exclusive & engrossing selfishness which accompanies it." *CW*, vol. 12, *Earlier Letters*, 31.
64. *CW*, vol. 1, *Autobiography and Literary Essays*, 185.
65. See Morison, *John Austin*, 20–21.
66. For an excellent treatment of Austin's influence on Mill's series of articles titled *The Spirit of the Age*, see Richard B. Friedman, "An Introduction to Mill's Theory of Authority," in *Mill: A Collection of Critical Essays*, ed. J.B. Schneewind (London: University of Notre Dame Press, 1969), 379–425.
67. John Austin, *The Province of Jurisprudence Determined; and, The Uses of the Study of Jurisprudence* (New York: Noonday Press, 1954), 73.
68. Austin, *Province of Jurisprudence Determined*, 79.
69. For d'Eichtal's impression of Mill on this occasion, see his diary entry, printed in "Condition de la classe ouvriere en Angleterre (1828)," *Revue Historique* 79 (1902), 84.
70. Comte would have much more to say about his *science positive*, just as Mill would have much more to say about Comte's achievements and limitations; moreover, the two men would carry on an important correspondence during the 1840s.
71. *CW*, vol. 12, *Earlier Letters*, 34–38.
72. *CW*, vol. 12, *Earlier Letters*, 40.
73. Iris Wessel Mueller, *John Stuart Mill and French Thought* (Urbana, Illinois: University of Illinois Press, 1956), 69–70.
74. *CW*, vol. 12, *Earlier Letters*, 38–43.
75. *CW*, vol. 12, *Earlier Letters*, 44–49.
76. *CW*, vol. 12, *Earlier Letters*, 48.
77. *CW*, vol. 12, *Earlier Letters*, 45–46.
78. For Mill's speech, see *CW*, vol. 26, *Journals and Debating Speeches*, 443–53.
79. Mill refers to Sterling's departure from the London Debating Society in the Early Draft of the *Autobiography*. See *CW*, vol. 1, *Autobiography and Literary Essays*, 162: "One vehement encounter between Sterling and me, he making what I thought a violent and unfair attack on the political philosophy I professed, to which I responded as sharply, fixed itself particularly in my memory because it was immediately followed by two things: one was, Sterling's withdrawing from the society; the other, that he and I sought one another privately much more than before, and became very intimate."
80. *CW*, vol. 12, *Earlier Letters*, 29.

Notes

81. *CW*, vol. 12, *Earlier Letters*, 45.
82. *CW*, vol. 1, *Autobiography and Literary Essays*, 189.
83. Bain, *James Mill*, 334.
84. *CW*, vol. 1, *Autobiography and Literary Essays*, 189.
85. The visitor was J. Crompton, a classmate of James Bentham Mill's at the University of London. This typescript, kindly shown to me by Ann P. Robson, derives from notes taken by A.S. West of an April 1875 conversation with the Rev. J. Crompton, who was then living in Norwich. The original manuscript is located in the Library of King's College, Cambridge.
86. *CW*, vol. 1, *Autobiography and Literary Essays*, 189.
87. *CW*, vol. 1, *Autobiography and Literary Essays*, 169.
88. *CW*, vol. 1, *Autobiography and Literary Essays*, 169.
89. Quoted in J.H. Burns, "The Light of Reason: Philosophical History in the Two Mills," in *James and John Stuart Mill: Papers of the Centenary Conference*, ed. John M. Robson and Michael Laine (Toronto: University of Toronto Press, 1976), 16.
90. *Works and Correspondence of David Ricardo*, vol. 7, 195–96.
91. Quoted in Burns, "The Light of Reason," 17.
92. *CW*, vol. 1, *Autobiography and Literary Essays*, 11.
93. Quoted in Burns, "The Light of Reason," 14.
94. *CW*, vol. 12, *Earlier Letters*, 76.
95. Two such collections of James Mill's essays appeared, the first consisting of four articles, the second of seven. Neither volume provides a year of publication. Current thought is that the first collection appeared in 1823 and the second in 1825. See Jack Lively and John Rees, eds., *Utilitarian Logic and Politics* (Oxford: Oxford University Press), 52. Mill's essay is reprinted in *Utilitarian Logic and Politics*, 53–95.
96. *CW*, vol. 1, *Autobiography and Literary Essays*, 107.
97. Quoted in Stefan Collini, Donald Winch, and John Burrow, *That Noble Science of Politics: A Study in Nineteenth-Century Intellectual History* (Cambridge: Cambridge University Press, 1983), 100. Leslie Stephen aptly stated that James Mill, in his *Essay on Government*, "speaks as from the chair of a professor laying down the elementary principles of a demonstrated science." Leslie Stephen, *The English Utilitarians*, 3 vols. (London: Duckworth, 1900), vol. 2, 75.
98. *Essay on Government*, in *Utilitarian Logic and Politics*, 57.
99. *Essay on Government*, in *Utilitarian Logic and Politics*, 79–80. The scope of James Mill's commitment to a democratic suffrage has been a subject of scholarly debate. Some historians maintain that for tactical reasons he refrained from advocating universal manhood suffrage, his true preference, because he knew the great bulk of propertied Englishmen strenuously opposed it. Others argue that he preferred a restricted suffrage, one that would place real power in the hands of the middle classes. The major contributions to this debate: Joseph Hamburger, "James Mill on Universal Suffrage and the Middle Class," *Journal of Politics* 24 (1962): 167–90; W.E.S. Thomas, "James Mill's Politics: The 'Essay on Government' and the Movement for Reform," *Historical Journal* 12 (1969): 249–84; W.R. Carr, "James Mill's Politics Reconsidered: Parliamentary Reform and the Triumph of Truth," *Historical Journal* 14 (1971): 553–80; W.E.S. Thomas, "James Mill's Politics: A Rejoinder," *Historical Journal* 14 (1971): 735–50; W.R. Carr, "James Mill's Politics: A Final Word," *Historical Journal*, 15 (1972): 315–20.
100. Clive, *Macaulay*, 47.
101. For the impression Macaulay made at the Union, see Clive, *Macaulay*, 46–50.
102. "The Present Administration," *Edinburgh Review* 46 (1827): 261.
103. "Mill's Essay on Government: Utilitarian Logic and Politics," *Edinburgh Review* 49 (1829): 160–61.
104. "The Present Administration," 262.
105. "The Present Administration," 263.

106. Lord Grey, the venerable Whig leader, detested Canning, and would not accept office in his administration.
107. "The Present Administration," 261.
108. For a stimulating treatment of James Mill's attempts to exploit this fear in the interest of the reform cause, see Joseph Hamburger, *James Mill and the Art of Revolution* (New Haven, Connecticut: Yale University Press, 1963).
109. The friendship of James Mill and Henry Brougham, both graduates of the University of Edinburgh, lasted from the early nineteenth century until Mill's death in 1836. In personality they were polar opposites, Brougham being emotionally demonstrative, and ready for others to respond in kind. James Mill seems to have felt a real affection for him, and certainly valued his political services. In 1833, at a time when Brougham, as Lord Chancellor, was undertaking sweeping reforms of the legal system in a Benthamite spirit, he got appreciative and encouraging words from James Mill: "I hope you consider one duty, the care of your health. I know not when the time was, in the history of our species, that more depended on the health of one man, than depends at this moment on yours. The progress of mankind would lose a century by the loss of you." Bain, *James Mill*, 371. Brougham would have heartily concurred in this estimate of his own importance.
110. "The London University," *Edinburgh Review* 43 (1826): 315–41. It has been forcefully argued, in fact, that Macaulay was drawn to certain aspects of Utilitarian thought; see William Thomas, *The Quarrel of Macaulay and Croker: Politics and History in the Age of Reform* (Oxford: Oxford University Press, 2000), 70–73.
111. "Periodical Literature: *Edinburgh Review*," *Westminster Review* 1 (1824): 206–49 and 505–41.
112. *CW*, vol. 1, *Autobiography and Literary Essays*, 312–13.
113. "Mill's Essay on Government: Utilitarian Logic and Politics," 161, 162.
114. "Mill's Essay on Government: Utilitarian Logic and Politics," 185–86.
115. "Mill's Essay on Government: Utilitarian Logic and Politics," 188–89.
116. *CW*, vol. 1, *Autobiography and Literary Essays*, 165.
117. *CW*, vol. 1, *Autobiography and Literary Essays*, 165, 167.
118. *CW*, vol. 7, *System of Logic*, 452.
119. *CW*, vol. 18, *Essays on Politics and Society*, 22.
120. Quoted in Collini, Winch, and Burrow, *That Noble Science of Politics*, 103.
121. *CW*, vol. 7, *System of Logic*, 603.
122. *CW*, vol. 1, *Autobiography and Literary Essays*, 189.
123. *CW*, vol. 1, *Autobiography and Literary Essays*, 177.
124. *CW*, vol. 12, *Earlier Letters*, 29–30.
125. Tooke had acted as the initial conduit between the Saint-Simonians and Mill. After dining with d'Eichtal on 30 May 1828, Tooke led his Parisian guest to the meeting of the London Debating Society at which d'Eichtal witnessed Mill's powerful mind in action.
126. *CW*, vol. 12, *Earlier Letters*, 44–45.
127. *CW*, vol. 1, *Autobiography and Literary Essays*, 83.
128. *CW*, vol. 32, *Additional Letters*, 9.
129. *CW*, vol. 12, *Earlier Letters*, 245. What did Mill make of the demons that drove Eyton Tooke to take his own life? In his letter to Nichol, he attributed his friend's "malady" to "intense and unremitting study." On the specific nature of the "malady" itself he did not elaborate. More than six decades after the event a book appeared that offered an explanation of Tooke's suicide. Henry Solly, in his memoir *These Eighty Years*, asserted that Tooke had fallen desperately in love with Solly's sister and had killed himself in the mistaken belief that she had rejected him (*These Eighty Years*, vol. 1, 134–38). We have no means of knowing whether Mill was aware of this turbulent emotional entanglement. Nor, of course, can we know how large a role it played in the tragic outcome. The letter Mill wrote Nichol in 1834 implied that Mill's personal inquest into the death of Eyton Tooke had issued in a verdict of "Broken Mind" rather than "Broken Heart."

Three Mill and Harriet Taylor: The Early Years

1. This oration was given while Smith stood before Bentham's corpse, which lay on the dissecting table, Bentham having bequeathed his body to science.
2. Quoted in Hayek, *John Stuart Mill and Harriet Taylor*, 25. Harriet Taylor's slight build may have made her appear taller than she was; her passport, issued in 1838, listed her height at 5 feet 1 inch (see Mill-Taylor Collection, British Library of Political and Economic Science, Box III/68).
3. Jacobs, ed., *Complete Works of Harriet Taylor Mill*, 440.
4. Jacobs, ed., *Complete Works of Harriet Taylor Mill*, 437.
5. Bain, *John Stuart Mill*, 164n.
6. Bain, *John Stuart Mill*, 164n.
7. *Collected Letters of Thomas and Jane Welsh Carlyle*, vol. 7, ed. Charles Richard Sanders (Durham, North Carolina: Duke University Press, 1977), 269.
8. See Ann P. Robson and John M. Robson, eds., *Sexual Equality: Writings by John Stuart Mill, Harriet Taylor, and Helen Taylor* (Toronto: University of Toronto Press, 1994), xvi.
9. See R.K. Webb, "William Johnson Fox (1786–1864)," in *The Oxford Dictionary of National Biography*, ed. H.C.G. Matthew and Brian Harrison (Oxford: Oxford University Press, 2004), vol. 20, 690.
10. We lack firm evidence for the date of this fateful dinner. The year we know from what Mill says in the *Autobiography*—"My first introduction to the lady who, after a friendship of twenty years, consented to become my wife, was in 1830" (*CW*, vol. 1, *Autobiography and Literary Essays*, 193), and from a letter written by Harriet Taylor Mill to her second husband in 1854, when the first draft of the *Autobiography* was under construction—"Should there not be a summary of *our* relationship from its commencement in 1830." Jacobs, ed., *Complete Works of Harriet Taylor Mill*, 375.
11. *Collected Letters of Thomas and Jane Welsh Carlyle*, vol. 7, 245–46.
12. All this by way of offering some kind of rational explanation for that unalterably nonrational phenomenon—romantic love.
13. *CW*, vol. 1, *Autobiography and Literary Essays*, 195.
14. *CW*, vol. 23, *Newspaper Writings*, 436.
15. Moncure Daniel Conway, *Centenary History of the South Place Society* (London: Williams and Norgate, 1894), 89.
16. For Conway's obituary of Mill, see *Harper's Monthly Magazine* 47 (1873), 528–34.
17. In contending that there are ample grounds for skepticism regarding the tale of a marriage proposal, I do not mean to suggest that Mill was incapable of developing a romantic attachment to anyone other than Harriet Taylor, either before or after 1830. His feelings for Eliza Flower very probably included sentiments of a romantic nature; the circumstances of the case, however, do not fit the notion of his having acted on those sentiments. Yet another woman—another Harriet yet—has been mentioned in connection with Mill's susceptibilities in this sphere: Lady Harriet Baring (wife of William Bingham Baring, second Lord Ashburton). Well known is Charles Greville's allusion to this connection in the immediate aftermath of Lady Harriet's death in 1857. "Two men were certainly in love with her, both distinguished in different ways. One was John Mill, who was sentimentally attached to her, and for a long time was devoted to her society. She was pleased and flattered by his devotion, but as she did not in the slightest degree return his passion, though she admired his abilities, he at last came to resent her indifference; and ended by estranging himself from her entirely, and proved the strength of his feeling by his obstinate refusal to continue even his acquaintance with her." (*Greville Memoirs*, vol. 2, 52; the second man named by Greville was Charles Buller.) Greville tends to be a reliable source, in the sense that he was uncommonly well-informed and did not

write what he knew or suspected to be untrue. He himself had known Lady Harriet fairly well, as he had Charles Buller. Of course Greville's informant could have been given to wild exaggeration. Packe accepts that Mill knew her, but supposes that any romantic feelings on Mill's side could not have been held after 1830. "Although the tradition is that Charles Buller introduced Mill to Lady Harriet Baring, and Charles Buller did not himself meet her until 1837, it is hard to believe that Mill can have behaved in the manner described by Greville so long after the events of the summer of 1830." *Life of John Stuart Mill*, 110n. In the seven volumes of Mill's printed letters, Harriet Baring's name does not appear. Yet there is evidence of a correspondence between them in the 1840s. An editorial footnote in the Carlyles' *Collected Letters* quotes a letter from Mill to Lady Harriet Baring dated August 1843 (the passage in question praises John Sterling's recently published *Strafford: A Tragedy*). The location given for this manuscript letter is "MS: Marquess of Northampton." See *Collected Letters of Thomas and Jane Welsh Carlyle*, vol. 17, ed. Clyde de L. Ryals and Kenneth J. Fielding (Durham, North Carolina: Duke University Press, 1990), 42n. A letter from Jane Carlyle to her husband, dated 31 July 1843, also links Mill and Lady Harriet Baring. According to Mrs. Carlyle, Mill had arranged for Giuseppe Mazzini to meet Lady Harriet Baring. She reported to her husband, whose own infatuation with Lady Harriet caused Jane Carlyle no end of misery, that "Mazzini's visit to Lady Baring (as he calls her) went off wonderfully well. I am afraid my dear this Lady Baring of yours and his and John Mill's and everybody's is an arch coquette"; she attributed to Mazzini the words "John Mill appeared to be *loving* her very much, and taking great pains to show her that his opinions were the right ones." *Collected Letters of Thomas and Jane Welsh Carlyle*, vol. 16, ed. Clyde de L. Ryals and Kenneth J. Fielding (Durham, North Carolina: Duke University Press, 1990), 329. A number of highly intelligent men found Lady Harriet Baring's company intoxicating. I do not have a hard time believing that Mill was one of them.
18. Norton made a verbatim record of this conversation, which is included in Sarah Norton and M.A. Dewolfe Howe, eds., *The Letters of Charles Eliot Norton*, 2 vols. (Boston: Houghton Mifflin, 1913), vol. 1, 496–97.
19. Leader, ed., *Life and Letters of John Arthur Roebuck*, 38.
20. *CW*, vol. 12, *Earlier Letters*, 114.
21. *CW*, vol. 12, *Earlier Letters*, 158.
22. *CW*, vol. 12, *Earlier Letters*, 159.
23. Jacobs, ed., *Complete Works of Harriet Taylor Mill*, 323–24.
24. The paper on which Mill wrote his essay carries a watermark of 1831; Harriet Taylor also conveyed her thoughts to Mill on the subject, on paper watermarked 1832. For a discussion of the rationale for dating these compositions to late 1832 or early 1833, see John M. Robson's textual introduction to *CW*, vol. 21, *Essays on Equality, Law, and Education*, ed. John M. Robson (Toronto: University of Toronto Press, 1984), lviii–lx.
25. *CW*, vol. 21, *Essays on Equality, Law, and Education*, 37. The essay opens with the words quoted.
26. *CW*, vol. 1, *Autobiography and Literary Essays*, 253n.
27. Jeremy Bentham, *An Introduction to the Principles of Morals and Legislation*, ed. J.H. Burns and H.L.A. Hart (London: Athlone, 1970), 245n. Among notes Bentham wrote in response to James Mill's *Essay on Government* is one that alludes to "the already universally existing tyranny of the male sex over the female." See Bhikhu Parekh, ed., *Bentham's Political Thought* (New York: Barnes & Noble, 1973), 312.
28. A letter John Mill wrote to Sir Samuel Bentham in 1819 reported that his sister Wilhelmina, then eleven years old, had read "some Cæsar; almost all Phædrus, all the Catiline and part of the Jugurtha of Sallust, and is now reading the Eclogues of Virgil." As for his sister Clara, then nine years old, she "had begun Latin also. After going through the grammar, she read some Cornelius Nepos and Cæsar, almost as much as Willie of Sallust, and is now reading Ovid. They are both tolerably good arithmeticians; they have gone so far as the extraction of the

cube root. They are reading the Roman Antiquities and the Greek Mythology, and are translating English into Latin from Mair's Introduction to Latin Syntax." *CW*, vol. 12, *Earlier Letters*, 10.
29. *CW*, vol. 1, *Autobiography and Literary Essays*, 312.
30. *CW*, vol. 1, *Autobiography and Literary Essays*, 107.
31. *CW*, vol. 1, *Autobiography and Literary Essays*, 253n.
32. *CW*, vol. 21, *Essays on Equality, Law, and Education*, 40.
33. *CW*, vol. 21, *Essays on Equality, Law, and Education*, 41.
34. *CW*, vol. 21, *Essays on Equality, Law, and Education*, 42.
35. *CW*, vol. 21, *Essays on Equality, Law, and Education*, 42.
36. *CW*, vol. 21, *Essays on Equality, Law, and Education*, 43.
37. *CW*, vol. 21, *Essays on Equality, Law, and Education*, 44.
38. *CW*, vol. 21, *Essays on Equality, Law, and Education*, 44.
39. *CW*, vol. 21, *Essays on Equality, Law, and Education*, 44–45.
40. *CW*, vol. 21, *Essays on Equality, Law, and Education*, 45.
41. *CW*, vol. 21, *Essays on Equality, Law, and Education*, 46.
42. *CW*, vol. 21, *Essays on Equality, Law, and Education*, 46–47.
43. *CW*, vol. 21, *Essays on Equality, Law, and Education*, 47.
44. *CW*, vol. 21, *Essays on Equality, Law, and Education*, 48.
45. *CW*, vol. 21, *Essays on Equality, Law, and Education*, 48.
46. *CW*, vol. 21, *Essays on Equality, Law, and Education*, 49.
47. *CW*, vol. 21, *Essays on Equality, Law, and Education*, 39.
48. *CW*, vol. 21, *Essays on Equality, Law, and Education*, Appendix A, 375.
49. *CW*, vol. 21, *Essays on Equality, Law, and Education*, Appendix A, 375–76.
50. *CW*, vol. 21, *Essays on Equality, Law, and Education*, Appendix A, 376.
51. *CW*, vol. 21, *Essays on Equality, Law, and Education*, Appendix A, 376.
52. *CW*, vol. 21, *Essays on Equality, Law, and Education*, Appendix A, 376.
53. *CW*, vol. 21, *Essays on Equality, Law, and Education*, Appendix A, 377.
54. *CW*, vol. 21, *Essays on Equality, Law, and Education*, Appendix A, 377.
55. *CW*, vol. 21, *Essays on Equality, Law, and Education*, Appendix A, 377.
56. *CW*, vol. 12, *Earlier Letters*, 187.
57. *CW*, vol. 12, *Earlier Letters*, 144.
58. *CW*, vol. 12, *Earlier Letters*, 144.
59. *CW*, vol. 12, *Earlier Letters*, 174–75.
60. Jacobs, ed., *Complete Works of Harriet Taylor Mill*, 327–28.
61. Jacobs, ed., *Complete Works of Harriet Taylor Mill*, 331.
62. *CW*, vol. 12, *Earlier Letters*, 187.
63. In a letter to Carlyle of 5 October 1833, Mill noted that he was "going to Paris probably at the end of the week." He expected his stay to last five weeks (as much time as his autumn vacation from India House would permit). See *CW*, vol. 12, *Earlier Letters*, 180.
64. *CW*, vol. 12, *Earlier Letters*, 187.
65. *CW*, vol. 12, *Earlier Letters*, 185–86.
66. *CW*, vol. 12, *Earlier Letters*, 187.
67. *CW*, vol. 12, *Earlier Letters*, 187–88.
68. *CW*, vol. 12, *Earlier Letters*, 189.
69. *CW*, vol. 12, *Earlier Letters*, 188.
70. Jacobs, ed., *Complete Works of Harriet Taylor Mill*, 329.
71. Jacobs, ed., *Complete Works of Harriet Taylor Mill*, 330.
72. Jacobs, ed., *Complete Works of Harriet Taylor Mill*, 330.
73. *CW*, vol. 12, *Earlier Letters*, 215.
74. *CW*, vol. 12, *Earlier Letters*, 213–14.
75. *Collected Letters of Thomas and Jane Welsh Carlyle*, vol. 7, 174.

76. Sarah Austin was at this time carrying on a passionate correspondence with Prince Hermann Puckler-Muskau (Prince Pickling Mustard to certain London wags); perhaps one of several reasons for her keen interest in the Mill-Taylor story.
77. Leader, ed., *Life and Letters of John Arthur Roebuck*, 38.
78. *CW*, vol. 1, *Autobiography and Literary Essays*, 236.
79. Quoted in Sarah Wilks, "The Mill-Roebuck Quarrel," *Mill News Letter* 13 (Summer, 1978): 10.
80. *CW*, vol. 12, *Earlier Letters*, 227.
81. See Francis E. Mineka, *The Dissidence of Dissent: The Monthly Repository* (Chapel Hill, North Carolina: University of North Carolina Press, 1944), 188–97.
82. *CW*, vol. 12, *Earlier Letters*, 227–28.
83. *CW*, vol. 12, *Earlier Letters*, 227n.
84. Jacobs, ed., *Complete Works of Harriet Taylor Mill*, 330–32.
85. Jacobs, ed., *Complete Works of Harriet Taylor Mill*, 332–33.
86. Jacobs, ed., *Complete Works of Harriet Taylor Mill*, 331.
87. *CW*, vol. 1, *Autobiography and Literary Essays*, 199.
88. *CW*, vol. 1, *Autobiography and Literary Essays*, 198.
89. *CW*, vol. 1, *Autobiography and Literary Essays*, 198.
90. Harriet Taylor's careful reading of this passage probably had something to do with its removal from the text. It did not appear in the final version of Mill's *Autobiography*.

Four   Mystifying the Mystic: Mill and Carlyle in the 1830s

1. The best modern biography is Kaplan's *Thomas Carlyle*.
2. See D.A. Haury, *The Origins of the Liberal Party and Liberal Imperialism: The Career of Charles Buller, 1806–1848* (New York: Garland, 1987).
3. Not until the summer of 1825 did Carlyle learn from Jane Welsh that she had "*once* passionately loved" Irving; see Kaplan, *Thomas Carlyle*, 114.
4. Carlyle had become fed up with Mrs. Buller, whose treatment of him he thought inconsiderate. She apparently had a hard time figuring out where she wanted to reside. In mid-1824 the choice lay between London, Cornwall, and Boulogne. She eventually opted for the last, and Carlyle did not find this prospect palatable. For one thing, it would render more difficult his pursuit of Jane Welsh, with whom he shared his relief at being freed from the Bullers, in a letter dated 22 July 1824. "We parted good friends, as I positively declined accompanying them to France, and myself advised the sending of their son to Cambridge immediately. Mrs Buller cannot now shift my residence and plague me with the consequences of her ignorant caprice from week to week; the dead-hearted fashionable frivolity of her and hers are now their own concern: I am a free man; let any arid, gaudy, drivelling male or female dandy come within wind of me to fret me and disgust me with their inanities if they dare!" *Collected Letters of Thomas and Jane Welsh Carlyle*, vol. 3, ed. Charles Richard Sanders (Durham, North Carolina: Duke University Press, 1970), 114.
5. *Collected Letters of Thomas and Jane Welsh Carlyle*, vol. 1, ed. Charles Richard Sanders (Durham, North Carolina: Duke University Press, 1970), 389.
6. Quoted in C.F. Harrold, *Carlyle and German Thought* (New Haven, Connecticut: Yale University Press, 1934), 184.
7. Quoted in G.B. Tennyson, *Sartor Called Resartus: The Genesis, Structure, and Style of Thomas Carlyle's First Major Work* (Princeton, New Jersey: Princeton University Press, 1965), 91. The synopsis presented in this paragraph draws heavily on Tennyson's superb study.

8. *Collected Letters of Thomas and Jane Welsh Carlyle*, vol. 5, 235n.
9. "Signs of the Times," *Edinburgh Review* 49 (1829): 458–59.
10. *Collected Letters of Thomas and Jane Welsh Carlyle*, vol. 4, ed. Charles Richard Sanders (Durham, North Carolina: Duke University Press, 1970), 335.
11. *Collected Letters of Thomas and Jane Welsh Carlyle*, vol. 4, 390.
12. "Characteristics," *Edinburgh Review* 54 (1831): 382.
13. In 1847 Empson succeeded Macvey Napier as editor of the *Edinburgh Review*, a position he held until his sudden death from influenza in 1852.
14. *Collected Letters of Thomas and Jane Welsh Carlyle*, vol. 5, 379.
15. *Collected Letters of Thomas and Jane Welsh Carlyle*, vol. 5, 398.
16. *Collected Letters of Thomas and Jane Welsh Carlyle*, vol. 5, 428.
17. *CW*, vol. 12, *Earlier Letters*, 85–86.
18. *CW*, vol. 12, *Earlier Letters*, 101.
19. *CW*, vol. 1, *Autobiography and Literary Essays*, 163.
20. *Collected Letters of Thomas and Jane Welsh Carlyle*, vol. 6, ed. Charles Richard Sanders (Durham, North Carolina: Duke University Press, 1977), 196.
21. *CW*, vol. 12, *Earlier Letters*, 113.
22. *CW*, vol. 12, *Earlier Letters*, 111.
23. *Collected Letters of Thomas and Jane Welsh Carlyle*, vol. 6, 237.
24. *CW*, vol. 12, *Earlier Letters*, 128.
25. *CW*, vol. 12, *Earlier Letters*, 143–44.
26. *CW*, vol. 1, *Autobiography and Literary Essays*, 329; for Carlyle's treatment of this idea, see "Characteristics," *Edinburgh Review* 54 (1831): 357–63, 367–68, 372–73.
27. Adams became the husband of Sarah Flower in 1834.
28. *CW*, vol. 1, *Autobiography and Literary Essays*, 369–70.
29. *Collected Letters of Thomas and Jane Welsh Carlyle*, vol. 6, 377.
30. *CW*, vol. 12, *Earlier Letters*, 153–54.
31. *CW*, vol. 12, *Earlier Letters*, 154.
32. *Collected Letters of Thomas and Jane Welsh Carlyle*, vol. 6, 400–01.
33. *CW*, vol. 12, *Earlier Letters*, 161.
34. *CW*, vol. 12, *Earlier Letters*, 163. Mill's first formal statement of the relation between science and art appeared in his essay "On the Definition of Political Economy," which he published in the *London and Westminster Review* in 1836. The initial draft of this essay was probably written in the autumn of 1831, and then revised during the summer of 1833. Mill reprinted the piece in 1844, as the concluding item of his *Essays on Some Unsettled Questions of Political Economy*. For his discussion in the essay of "the line of separation . . . between science and art," see *CW*, vols. 4–5, *Essays on Economics and Society*, ed. John M. Robson (Toronto: University of Toronto Press, 1967), vol. 4, 331n.
35. See *CW*, vol. 7, *System of Logic*, textual introduction, lviii.
36. *Collected Letters of Thomas and Jane Welsh Carlyle*, vol. 7, 8. Carlyle added, however, that Mill was "much too exclusively *logical*. I think, he will mend: but his character is naturally not *large*, rather high and solid."
37. *Collected Letters of Thomas and Jane Welsh Carlyle*, vol. 6, 437.
38. *Collected Letters of Thomas and Jane Welsh Carlyle*, vol. 6, 439.
39. *CW*, vol. 12, *Earlier Letters*, 190–97.
40. *Collected Letters of Thomas and Jane Welsh Carlyle*, vol. 7, 54–55.
41. *CW*, vol. 12, *Earlier Letters*, 204–05.
42. *CW*, vol. 12, *Earlier Letters*, 204.
43. *CW*, vol. 12, *Earlier Letters*, 206.
44. *CW*, vol. 12, *Earlier Letters*, 207–08.
45. *Collected Letters of Thomas and Jane Welsh Carlyle*, vol. 7, 72–73.
46. *CW*, vol. 1, *Autobiography and Literary Essays*, 183.

47. *CW*, vol. 12, *Earlier Letters*, 219.
48. *CW*, vol. 12, *Earlier Letters*, 152.
49. Edward Lytton Bulwer, *England and the English*, 2 vols. (London: Bentley, 1833), vol. 2, Appendix B, 321–44.
50. *CW*, vol. 1, *Autobiography and Literary Essays*, 207.
51. *CW*, vol. 10, *Essays on Ethics, Religion and Society*, 12.
52. *CW*, vol. 10, *Essays on Ethics, Religion and Society*, 15.
53. *CW*, vol. 12, *Earlier Letters*, 172.
54. *CW*, vol. 12, *Earlier Letters*, 236.
55. *CW*, vol. 1, *Autobiography and Literary Essays*, Appendix G, 613.
56. *CW*, vol. 12, *Earlier Letters*, 175.
57. Jacobs, ed., *Complete Works of Harriet Taylor Mill*, 327.
58. *CW*, vol. 1, *Autobiography and Literary Essays*, Appendix G, 612.
59. *Collected Letters of Thomas and Jane Welsh Carlyle*, vol. 6, 446–47.
60. *CW*, vol. 12, *Earlier Letters*, 184.
61. *Collected Letters of Thomas and Jane Welsh Carlyle*, vol. 7, 152.
62. *Collected Letters of Thomas and Jane Welsh Carlyle*, vol. 7, 327.
63. *Collected Letters of Thomas and Jane Welsh Carlyle*, vol. 7, 174.
64. *Collected Letters of Thomas and Jane Welsh Carlyle*, vol. 7, 245–46.
65. *Collected Letters of Thomas and Jane Welsh Carlyle*, vol. 7, 259–60.
66. Carlyle wrote: "We are to dine there on Tuesday, and meet a new set of persons, said, among other qualities, to be interested in *me*. The Editor of Fox's Repository (Fox himself) is the main man I care for." See *Collected Letters of Thomas and Jane Welsh Carlyle*, vol. 7, 260.
67. *Collected Letters of Thomas and Jane Welsh Carlyle*, vol. 7, 269–70.
68. *Collected Letters of Thomas and Jane Welsh Carlyle*, vol. 7, 326–27. The Carlyles' fascination with Harriet Taylor did not end in 1834; in early 1835 Jane Carlyle, in a postscript to a letter her husband wrote his brother John, archly commented: "There is a *Mrs. Tailor* [*sic*] whom I could really love; if it were safe and she were willing—but she is a dangerous looking woman and engrossed with a dangerous passion and no useful relation can spring up between us." *Collected Letters of Thomas and Jane Welsh Carlyle*, vol. 8, ed. Charles Richard Sanders (Durham, North Carolina: Duke University Press, 1981), 15.
69. *CW*, vol. 1, *Autobiography and Literary Essays*, 65.
70. *CW*, vol. 20, *Essays on French History and Historians*, ed. John M. Robson (Toronto: University of Toronto Press, 1985), 58.
71. *CW*, vol. 20, *Essays on French History and Historians*, 4–5.
72. *CW*, vol. 20, *Essays on French History and Historians*, 99.
73. For a concise and illuminating discussion of Carlyle's conception of history, see A.L. Le Quesne, *Carlyle* (Oxford: Oxford University Press, 1982), 33–35; for a valuable extended treatment of this subject, see J.D. Rosenberg, *Carlyle and the Burden of History* (Cambridge, Massachusetts: Harvard University Press, 1985).
74. *Collected Letters of Thomas and Jane Welsh Carlyle*, vol. 7, 24.
75. *Collected Letters of Thomas and Jane Welsh Carlyle*, vol. 6, 446.
76. *CW*, vol. 12, *Earlier Letters*, 181–82.
77. *CW*, vol. 1, *Autobiography and Literary Essays*, 45.
78. *Collected Letters of Thomas and Jane Welsh Carlyle*, vol. 7, 236.
79. *Collected Letters of Thomas and Jane Welsh Carlyle*, vol. 8, 12.
80. *Collected Letters of Thomas and Jane Welsh Carlyle*, vol. 8, 53.
81. Portion of Carlyle's Journal entry for 7 March 1835; see *Collected Letters of Thomas and Jane Welsh Carlyle*, vol. 8, 67n.
82. *Collected Letters of Thomas and Jane Welsh Carlyle*, vol. 8, 76.
83. *CW*, vol. 12, *Earlier Letters*, 253.
84. Packe, *Life of John Stuart Mill*, 151n.

Notes

85. *Collected Letters of Thomas and Jane Welsh Carlyle*, vol. 10, ed. Charles Richard Sanders (Durham, North Carolina: Duke University Press, 1985), 47–48.
86. See *Letters of Charles Eliot Norton*, vol. 1, 496.
87. *Collected Letters of Thomas and Jane Welsh Carlyle*, vol. 8, 68n.
88. For an account of the exchange between Harriet Isabella Mill and Carlyle, see *CW*, vol. 12, *Earlier Letters*, 252n.
89. *Collected Letters of Thomas and Jane Welsh Carlyle*, vol. 8, 68n.
90. Quoted in *Collected Letters of Thomas and Jane Welsh Carlyle*, vol. 8, 68n.
91. *CW*, vol. 12, *Earlier Letters*, 253. For Carlyle's offer, see *Collected Letters of Thomas and Jane Welsh Carlyle*, vol. 8, 72. He kindly observed: "I think of all men living you are henceforth the least likely to commit such an oversight again."
92. *Collected Letters of Thomas and Jane Welsh Carlyle*, vol. 8, 70, 71.
93. *CW*, vol. 12, *Earlier Letters*, 252.
94. *Collected Letters of Thomas and Jane Welsh Carlyle*, vol. 8, 74, 74n.
95. *CW*, vol. 12, *Earlier Letters*, 257.
96. *CW*, vol. 20, *Essays on French History and Historians*, 133.
97. Long after this friendship ended Carlyle fondly recalled the assistance Mill provided and the value Carlyle placed on this assistance. "Mill was very useful about 'French Revolution;' lent me all his books, which were quite a collection on that subject; gave me, frankly, clearly, and with zeal, all his better knowledge than my own (which was pretty frequently of use in this or the other detail); being full of eagerness for such an advocate in that cause as he felt I should be." See Carlyle's *Reminiscences*, ed. James Anthony Froude (New York: Charles Scribner's Sons, 1881), 409–10.
98. This review forms part of *CW*, vol. 20, *Essays on French History and Historians*, 131–66.
99. *Collected Letters of Thomas and Jane Welsh Carlyle*, vol. 9, ed. Charles Richard Sanders (Durham, North Carolina: Duke University Press, 1981), 255.
100. *CW*, vol. 13, *Earlier Letters*, 427.
101. *CW*, vol. 1, *Autobiography and Literary Essays*, 49.
102. Carlyle, *Sartor Resartus*, ed. C.F. Harrold (Garden City, New York: Doubleday, 1937), 266–67.
103. *CW*, vol. 13, *Earlier Letters*, 449.
104. *CW*, vol. 1, *Autobiography and Literary Essays*, 183.
105. *CW*, vol. 1, *Autobiography and Literary Essays*, 198.

Five   Mill and the Secret Ballot

1. See *CW*, vol. 19, *Essays on Politics and Society*, 311–39, especially 331–38. The section on the ballot in *Considerations on Representative Government*, also in *CW*, vol. 19, *Essays on Politics and Society*, 488–95, is, in large measure, drawn directly from *Thoughts on Parliamentary Reform*.
2. *CW*, vol. 19, *Essays on Politics and Society*, 332–33.
3. Joseph Hamburger, *Intellectuals in Politics: John Stuart Mill and the Philosophic Radicals* (New Haven: Yale University Press, 1965).
4. James Mill, *The History of British India*, 2nd ed., 6 vols. (London: Baldwin, Cradock, and Joy, 1820), vol. 3, 451–52.
5. *CW*, vol. 19, *Essays on Politics and Society*, 331n–32n.
6. James Mill, "The Ballot," *Westminster Review* 18 (1830): 1–39. For brief commentaries on this article, see William Thomas, *Philosophic Radicals*, 140–45, and Bruce L. Kinzer, *The Ballot Question in Nineteenth-Century English Politics* (New York: Garland, 1982), 11–12.
7. *Westminster Review* 18 (1830): 8.
8. Hamburger, *Intellectuals in Politics*, 30–75.

9. Hamburger, *Intellectuals in Politics*, 68–71.
10. All of these essays are included in *CW*, vol. 18, *Essays on Politics and Society*.
11. *CW*, vol. 18, *Essays on Politics and Society*, 25–26.
12. *CW*, vol. 18, *Essays on Politics and Society*, 26.
13. Mill had already referred to the effective use made by the Tories at the recent election of the illegitimate electoral influence in their possession under the system of open voting.
14. *CW*, vol. 6, *Essays on England, Ireland, and the Empire*, ed. John M. Robson (Toronto: University of Toronto Press, 1982), 300–01.
15. *CW*, vol. 12, *Earlier Letters*, 317.
16. See Kinzer, *Ballot Question*, 37.
17. *Parliamentary Debates*, 3rd ser., vol. 39 (1837): cols. 65–73.
18. *CW*, vol. 6, *Essays on England, Ireland, and the Empire*, 410.
19. *CW*, vol. 6, *Essays on England, Ireland, and the Empire*, 411.
20. *CW*, vol. 6, *Essays on England, Ireland, and the Empire*, 412.
21. *CW*, vol. 6, *Essays on England, Ireland, and the Empire*, 413.
22. *CW*, vol. 6, *Essays on England, Ireland, and the Empire*, 465–95.
23. *CW*, vol. 6, *Essays on England, Ireland, and the Empire*, 469.
24. *CW*, vol. 6, *Essays on England, Ireland, and the Empire*, 475–78.
25. *CW*, vol. 6, *Essays on England, Ireland, and the Empire*, 479.
26. *CW*, vol. 6, *Essays on England, Ireland, and the Empire*, 481.
27. *CW*, vol. 6, *Essays on England, Ireland, and the Empire*, 482.
28. *CW*, vol. 6, *Essays on England, Ireland, and the Empire*, 483.
29. The most widely read of the Chartist newspapers, the *Northern Star*, reflected this point of view shortly before the 1839 ballot debate. "The Ballot, so far from extending the franchise, would curtail it, inasmuch as property being the standard, the owners of property would take care not to arm a masked battery [the labourers] against themselves . . . Once pass the Ballot, and no more scrutiny into the acts of your trustees—no more deference by the trustee to popular opinion!" *Northern Star*, 15 June 1839. For a discussion of Chartist suspicion of middle-class radicals who gave high priority to the ballot, see Kinzer, *Ballot Question*, 47–50.
30. *CW*, vol. 6, *Essays on England, Ireland, and the Empire*, 481–82.
31. *CW*, vol. 13, *Earlier Letters*, 410.
32. See *Mill News Letter* 6 (Spring, 1971): 15–20. Discussing, in Book V of the *Logic*, the ambiguity in the common expression "influence of property," Mill observes that it "is sometimes used for the influence of respect for superior intelligence, or gratitude for the kind offices which persons of large property have it so much in their power to bestow; at other times for the influence of fear; fear of the worst sort of power, which large property also gives its possessor, the power of doing mischief to dependents." Mill then remarks that "to confound these two, is the standing fallacy of ambiguity brought against those who seek to purify the electoral system from corruption and intimidation." Following this statement in the first and second editions (1843 and 1846), but omitted from the third (1851) and subsequent editions, is a sentence in which Mill, when referring to the statement "The influence of property is beneficial," remarks: "granted; if the former species of influence and that alone be meant; but conclusions are thence drawn in condemnation of expedients which (like secret voting, for example,) would deprive property of some of its influence, though only of the latter and bad kind." See *CW*, vol. 8, *System of Logic*, 811, 811n. Robson seems fully justified in concluding that the omission of this sentence in 1851 can only be explained as resulting from a change of mind on the ballot.
33. For Mill's *Daily News* articles, see *CW*, vol. 25, *Newspaper Writings*, 1101–04, 1104–07, 1107–09.
34. *CW*, vol. 14, *Later Letters*, 103.
35. *CW*, vol. 1, *Autobiography and Literary Essays*, 261.
36. *CW*, vol. 14, *Later Letters*, 222.
37. Cobden was largely responsible for the formation of the Ballot Society in early 1853. See Kinzer, *Ballot Question*, 55–64.

38. *CW*, vol. 1, *Autobiography and Literary Essays*, 251, 253.
39. *CW*, vol. 14, *Later Letters*, 221.
40. *CW*, vol. 14, *Later Letters*, 218.
41. *CW*, vol. 19, *Essays on Politics and Society*, 331.
42. *CW*, vol. 19, *Essays on Politics and Society*, 331, 331n–32n.
43. *CW*, vol. 19, *Essays on Politics and Society*, 333.
44. *CW*, vol. 19, *Essays on Politics and Society*, 332.
45. *CW*, vol. 15, *Later Letters*, 607–08.
46. *CW*, vol. 15, *Later Letters*, 608.
47. *CW*, vol. 15, *Later Letters*, 558–59.
48. *CW*, vol. 19, *Essays on Politics and Society*, 335.
49. *CW*, vol. 19, *Essays on Politics and Society*, 337.
50. *CW*, vol. 19, *Essays on Politics and Society*, 333–34.
51. *CW*, vol. 19, *Essays on Politics and Society*, 336. For a suggestive analytical consideration of Mill's theoretical defense of public voting, see Urbinati, *Mill on Democracy*, 104–22.
52. For a discussion of the opposition to secret voting, see Bruce L. Kinzer, "The Un-Englishness of the Secret Ballot," *Albion* 10 (1978): 237–56.
53. *Report from the Select Committee on Parliamentary and Municipal Elections*, vol. 8, Q. and A. 6462 (1868–69).
54. Quoted in Kinzer, *Ballot Question*, 80–81.
55. Quoted in Kinzer, *Ballot Question*, 81.
56. David Cresap Moore, *The Politics of Deference: A Study of the Mid-Nineteenth Century English Political System* (Hassocks, Sussex: Harvester Press, 1976).
57. Moore, *Politics of Deference*, 425–27.
58. Moore, *Politics of Deference*, 405.
59. Moore, *Politics of Deference*, 442.
60. *Report from the Select Committee on Parliamentary and Municipal Elections*, vol. 8 (1868–69).
61. See Norman Gash, *Politics in the Age of Peel: A Study in the Technique of Parliamentary Representation 1830–1850* (London: Longman, 1953), especially 176–77.
62. T.J. Nossiter, *Influence, Opinion and Political Idioms in Reformed England: Case Studies from the North East 1832–1874* (Hassocks, Sussex: Harvester Press, 1975), 162.
63. Kinzer, *Ballot Question*, 57–58.
64. Charles Stuart Parker, *Life and Letters of Sir James Graham*, 2 vols. (London: J. Murray, 1907), vol. 2, 171.
65. Quoted in Kinzer, *Ballot Question*, 58.
66. *CW*, vol. 1, *Autobiography and Literary Essays*, 245.

Six   Mill and the Problem of Party

1. Henry Sidgwick, *The Elements of Politics*, 4th ed. (London: Macmillan, 1929), 590.
2. A.H. Birch, *Representative and Responsible Government: An Essay on the British Constitution* (Toronto: University of Toronto Press, 1964), 114.
3. Dennis F. Thompson, *John Stuart Mill and Representative Government* (Princeton, New Jersey: Princeton University Press, 1976), 118–21, 187.
4. Urbinati, *Mill on Democracy*, 102.
5. *CW*, vol. 19, *Essays on Politics and Society*, 455.
6. *CW*, vol. 19, *Essays on Politics and Society*, 456.
7. *CW*, vol. 19, *Essays on Politics and Society*, 373.
8. John M. Robson, *The Improvement of Mankind* (Toronto: University of Toronto Press, 1968), 191–99.

9. *CW*, vol. 19, *Essays on Politics and Society*, 348.
10. Hamburger, *Intellectuals in Politics*.
11. *CW*, vol. 6, *Essays on England, Ireland, and the Empire*, 165.
12. *CW*, vol. 1, *Autobiography and Literary Essays*, 315.
13. *CW*, vol. 6, *Essays on England, Ireland, and the Empire*, 467.
14. *CW*, vol. 10, *Essays on Ethics, Religion and Society*, 163.
15. Mill held that the Toryism of Peel's Conservative party, let alone that of Derby and Disraeli, had little in common with Coleridge's speculative Toryism. Although Mill recognized the obvious superiority of Peel to the mass of his followers, and was indeed prepared, in "Reorganization of the Reform Party," to regard Peel as the legitimate representative of Conservative principle in national politics, he did not have a high opinion of the Tory leader. In 1837 Mill wrote: "What gives Sir Robert Peel his personal influence? What makes so many adhere to him? The opinion, a greatly exaggerated one, entertained of his capacity for business . . . If Radicalism had its Sir Robert Peel, he would be at the head of an administration within two years: and Radicalism must be a barren soil if it cannot rival so sorry a growth as that; if it cannot produce a match for perhaps the least gifted man that ever headed a powerful party in this country . . . He does not know his age; he has always blundered miserably in his estimate of it. But he knows the House of Commons, and the sort of men of whom it is composed. He knows what will act upon their minds, and he is able to strike the right chord upon that instrument. He has, besides, all that the mere routine of office-experience can give, to a man who brought to it no principles drawn from a higher philosophy, and no desire for any." *CW*, vol. 6, *Essays on England, Ireland, and the Empire*, 404. It is perhaps ironic that in 1846 Disraeli's criticism of Peel resembled Mill's assessment of the Conservative leader, while Peel probably then shared Mill's view of the rank and file of the Tory party.
16. *CW*, vol. 15, *Later Letters*, 672.
17. *CW*, vol. 19, *Essays on Politics and Society*, 452n. Mill did not claim that all Conservatives were silly and stupid, only that the conduct of the party tended to reflect the silliness and stupidity characteristic of most of its members. "Is it not surprising that Conservatives have no sense or appreciation of Conservative principles? Conservatism with us means a blind opposition to change. I know no Conservatives who are really so but the Saturday reviewers whose adherence is to *principles* of stability & principles of unjust domination so far as now practically maintainable, but who have no mere instinctive attachment to details as they are." See *CW*, vol. 15, *Later Letters*, 667–68.
18. *CW*, vol. 28, *Public and Parliamentary Speeches*, 22.
19. *CW*, vol. 25, *Newspaper Writings*, 1101–04, 1104–07, 1107–09.
20. *CW*, vol. 25, *Newspaper Writings*, 1103–04.
21. *CW*, vol. 19, *Essays on Politics and Society*, 452n.
22. The idea that the political landscape of these years lacked sharp party definition was expressed by a number of well-informed men. In the autumn of 1855 Palmerston invited Lord Stanley, the son of the leader of the Conservative party, to join his ministry. Although Stanley declined the offer, he admitted, in his reply to Palmerston, that "of late years, the lines of demarcation which separate political parties have been finely drawn, and have even at times appeared to be altogether effaced." See J.R. Vincent, ed. *Disraeli, Derby and the Conservative Party: The Political Journals of Lord Stanley, 1849–69* (Hassocks, Sussex: Harvester Press, 1978), 138. In 1864 Lord Robert Cecil, the Tory politician and future prime minister (as Lord Salisbury), praised Palmerston's House of Commons for having "done that it is most difficult and most salutary for a Parliament to do—nothing." (This statement appears in an article Cecil wrote for the *Quarterly Review* 116 [1864]: 245.) In 1856 Gladstone wrote an article, "The Declining Efficiency of Parliament," in which he stated: "The interval between the two parties has, by the practical solution of so many contested questions, been very greatly narrowed." See *Quarterly Review* 99 (1856): 562.

23. *CW*, vol. 19, *Essays on Politics and Society*, 448–66.
24. *CW*, vol. 19, *Essays on Politics and Society*, 363.
25. For a treatment of the 1865 Westminster contest, see Bruce L. Kinzer, Ann P. Robson, and John M. Robson, *A Moralist In and Out of Parliament: John Stuart Mill at Westminster, 1865–1868* (Toronto: University of Toronto Press, 1992), 22–79.
26. See *CW*, vol. 16, *Later Letters*, 1005–07.
27. *CW*, vol. 28, *Public and Parliamentary Speeches*, 34.
28. *CW*, vol. 16, *Later Letters*, 1122.
29. *CW*, vol. 16, *Later Letters*, 1138.
30. *CW*, vol. 16, *Later letters*, 1197.
31. *CW*, vol. 1, *Autobiography and Literary Essays*, 275.
32. *CW*, vol. 1, *Autobiography and Literary Essays*, 276.
33. *CW*, vol. 28, *Public and Parliamentary Speeches*, 75–83.
34. *CW*, vol. 1, *Autobiography and Literary Essays*, 279.
35. *CW*, vol. 28, *Public and Parliamentary Speeches*, 58–68.
36. *CW*, vol. 28, *Public and Parliamentary Speeches*, 105–13, 151–62, 247–61.
37. *CW*, vol. 28, *Public and Parliamentary Speeches*, 323.
38. *CW*, vol. 28, *Public and Parliamentary Speeches*, 336.
39. *CW*, vol. 28, *Public and Parliamentary Speeches*, 361.
40. For treatment of the 1868 Westminster contest, see Kinzer, Robson, and Robson, *A Moralist In and Out of Parliament*, 218–68.
41. *CW*, vol. 16, *Later Letters*, 1463.
42. *CW*, vol. 16, *Later Letters*, 1488. The working men's candidates Mill had in mind were probably Edmond Beales, Charles Bradlaugh, George Howell, and W.R. Cremer, all of whom were defeated at this general election. The "University Liberals" were men who professed strong liberal opinions and had close ties to the old universities. Their political views in the 1860s were best exemplified in the articles that made up the volume, edited by Leslie Stephen, *Essays on Reform* (London: Macmillan, 1867). Those who can be so classified, and who unsuccessfully sought election in 1868, included G.C. Brodrick, E.A. Freeman, Auberon Herbert, George Young, Godfrey Lushington, and Charles Roundell. For an excellent study of the university liberals, see Christopher Harvie, *The Lights of Liberalism: Liberals and the Challenge of Democracy* (London: Allen Lane, 1976).
43. *CW*, vol. 16, *Later Letters*, 1493.
44. *CW*, vol. 17, *Later Letters*, 1547n–48n.
45. *CW*, vol. 28, *Public and Parliamentary Speeches*, 65.
46. *CW*, vol. 16, *Later Letters*, 1479.
47. *CW*, vol. 16, *Later Letters*, 1514.
48. *CW*, vol. 17, *Later Letters*, 1697.
49. *CW*, vol. 6, *Essays on England, Ireland, and the Empire*, 482–87.
50. *CW*, vol. 6, *Essays on England, Ireland, and the Empire*, 483.
51. This is not to deny that Mill assigned an important administrative and educative function to an intellectual elite, an elite that through its expertise, wide-ranging vision, and persuasive powers would secure for itself, in certain spheres, a measure of deference from all elements participating in the political system. But this was, at least in Mill's eyes, a nonclass elite, one whose authority would be acknowledged by both middle and working classes.
52. Thompson, *John Stuart Mill and Representative Government*.
53. See Michael Barker, *Gladstone and Radicalism: The Reconstruction of Liberal Policy in Britain, 1885–1894* (Hassocks, Sussex: Harvester Press, 1975); J.P. Parry, *Democracy & Religion: Gladstone and the Liberal Party 1867–1875* (Cambridge: Cambridge University Press, 1986); and Eugenio F. Biagini, *Liberty, Retrenchment and Reform: Popular Liberalism in the Age of Gladstone, 1860–1880* (Cambridge: Cambridge University Press, 1992).
54. *CW*, vol. 1, *Autobiography and Literary Essays*, 35.

Seven Mill and the Experience of Political Engagement

1. *CW*, vol. 19, *Essays on Politics and Society*, 327.
2. *CW*, vol. 28, *Public and Parliamentary Speeches*, 170.
3. *CW*, vol. 1, *Autobiography and Literary Essays*, 288.
4. *The Times*, 23 December 1868.
5. *CW*, vol. 1, *Autobiography and Literary Essays*, 278–79.
6. See Hamburger, *Intellectuals in Politics*; Thomas, *Philosophic Radicals*, 147–205; Ann P. Robson and John M. Robson, "Private and Public Goals: John Stuart Mill and the *London and Westminster*," in *Innovators and Preachers: The Role of the Editor in Victorian England*, ed. J.H. Wiener (Westport, Connecticut: Greenwood Press, 1985), 231–57.
7. See Mill's 1837 article "Parties and the Ministry," his 1838 articles "Radical Party and Canada: Lord Durham and the Canadians," "Lord Durham and His Assailants," and "Lord Durham's Return," and his 1839 article "Reorganization of the Reform Party," in *CW*, vol. 6, *Essays on England, Ireland, and the Empire*, 381–495.
8. Hamburger, *Intellectuals in Politics*.
9. In retrospect, Mill might have concluded that the material he had worked on was almost as intractable as that Disraeli claimed to have worked on in the late 1860s. Mill, however, would not concede that Disraeli genuinely made such an effort.
10. Mill's correspondence in the late 1830s shows a preoccupation with the politics of the day. See *CW*, vols. 12–13, *Earlier Letters*, 248–414, *passim*.
11. Most of the first draft of the *Logic* was finished before the publication of William Whewell's *Philosophy of the Inductive Sciences*, 2 vols. (London: Parker, 1840), whose helpfulness in the completion of the project Mill acknowledges in *CW*, vol. 1, *Autobiography and Literary Essays*, 231.
12. *CW*, vol. 13, *Earlier Letters*, 406.
13. *CW*, vol. 1, *Autobiography and Literary Essays*, 229.
14. *CW*, vol. 6, *Essays on England, Ireland, and the Empire*, 468–79.
15. *CW*, vol. 6, *Essays on England, Ireland, and the Empire*, 468.
16. *CW*, vol. 18, *Essays on Politics and Society*, 197–98.
17. *CW*, vol. 13, *Earlier Letters*, 483.
18. *CW*, vol. 13, *Earlier Letters*, 544.
19. *CW*, vol. 8, *System of Logic*, 926.
20. *CW*, vol. 8, *System of Logic*, 927.
21. For a superb discussion of Mill as public moralist, see Stefan Collini, *Public Moralists: Political Thought and Intellectual Life in Britain 1850–1930* (Oxford: Oxford University Press, 1991), 121–69.
22. The brief manuscript "What Is to Be Done with Ireland" and the pamphlet *England and Ireland*, *CW*, vol. 6, *Essays on England, Ireland, and the Empire*, 497–502, and 505–32, respectively.
23. *CW*, vol. 18, *Essays on Politics and Society*, 205–11.
24. See *CW*, vol. 24, *Newspaper Writings*, 879–1035.
25. For useful discussions of Mill's response to the famine, see Lynn Zastoupil, "Moral Government: J.S. Mill on Ireland," *Historical Journal* 16 (1983): 707–17, and Bruce L. Kinzer, *England's Disgrace? J.S. Mill and the Irish Question* (Toronto: University of Toronto Press, 2001), 44–86.
26. *CW*, vol. 13, *Earlier Letters*, 507.
27. *CW*, vol. 13, *Earlier Letters*, 544.
28. *CW*, vol. 13, *Earlier Letters*, 642.
29. *CW*, vol. 1, *Autobiography and Literary Essays*, 243.
30. *CW*, vol. 13, *Earlier Letters*, 710.

NOTES

31. *CW*, vol. 13, *Earlier Letters*, 711–15.
32. *CW*, vol. 13, *Earlier Letters*, 715.
33. *CW*, vol. 6, *Essays on England, Ireland, and the Empire*, 482.
34. *CW*, vol. 13, *Earlier Letters*, 533.
35. *CW*, vol. 25, *Newspaper Writings*, 1110–12; *CW*, vol. 20, *Essays on French History and Historians*, 317–63.
36. For an illuminating examination of Mill and the 1848 French Revolution, see J.C. Cairns's excellent introduction to *CW*, vol. 20, *Essays on French History and Historians*, lxxxiii–xci.
37. *CW*, vol. 25, *Newspaper Writings*, 1112.
38. For various perspectives on the impact of the events of 1848 and the relation between Chartism and middle-class radicalism, see Roland Quinault, "1848 and Parliamentary Reform," *Historical Journal* 31 (1988): 831–51; Margot Finn, *After Chartism: Class and Nation in English Radical Politics* (Cambridge: Cambridge University Press, 1993), chapter 2; and Miles Taylor, *The Decline of British Radicalism 1847–1860* (Oxford: Oxford University Press, 1995), 106–23.
39. *CW*, vol. 25, *Newspaper Writings*, 1101–04, 1104–07, 1107–09.
40. *CW*, vol. 25, *Newspaper Writings*, 1103.
41. *CW*, vol. 25, *Newspaper Writings*, 1109.
42. *CW*, vol. 25, *Newspaper Writings*, 1105–07.
43. *CW*, vol. 25, *Newspaper Writings*, 1105.
44. *CW*, vol. 14, *Later Letters*, 34.
45. *CW*, vol. 14, *Later Letters*, 45.
46. *CW*, vol. 1, *Autobiography and Literary Essays*, 245.
47. *CW*, vol. 1, *Autobiography and Literary Essays*, 235, 237.
48. *CW*, vol. 1, *Autobiography and Literary Essays*, 239.
49. *CW*, vol. 27, *Journals and Debating Speeches*, 662.
50. Mill's retirement at this time was prompted by the passage of the India Act of 1858, which, to his deep regret, dissolved the East India Company as an agency of government in India.
51. For an excellent examination of Helen Taylor's importance in Mill's life between 1858 and 1873, see Ann P. Robson, "Mill's Second Prize in the Lottery of Life," in *A Cultivated Mind*, ed. Laine, 215–41.
52. See Stefan Collini's splendid introduction to *CW*, vol. 21, *Essays on Equality, Law, and Education*, viii–xix.
53. The two volumes of *Dissertations and Discussions* included those previously published essays (now revised) that Mill considered of permanent value. As a commercial proposition, *Dissertations and Discussions* was rendered attractive to Mill and his publisher (Parker) by the great success enjoyed by *On Liberty*.
54. To Helen Taylor, Mill wrote: "The truth is that though I detest society for society's sake yet when I can do anything for the public objects I care about by seeing & talking with people I do not dislike it . . . I believe the little additional activity & change of excitement does me good, & that it is better for me to try to serve my opinions in other ways as well as with a pen in my hand. With such people as Hare and Fawcett it is a pleasure." *CW*, vol. 15, *Later Letters*, 675. For a valuable consideration of the circle that formed around Mill during this phase of his life, see Jeff Lipkes, *Politics, Religion and Classical Political Economy in Britain: John Stuart Mill and His Followers* (Houndmills, Basingstoke, Hampshire: Macmillan Press, 1999).
55. See chapter 5 above for my discussion of Mill and the ballot.
56. See *CW*, vol. 19, *Essays on Politics and Society*, 322–28.
57. *CW*, vol. 1, *Autobiography and Literary Essays*, 261.
58. This is the mode recommended in *Thoughts on Parliamentary Reform*. Mill found it set forth in James Garth Marshall, *Minorities and Majorities; Their Relative Rights. A Letter to Lord John Russell, M.P. on Parliamentary Reform* (London: Ridgway, 1853).
59. *CW*, vol. 15, *Later Letters*, 543–44.
60. *CW*, vol. 19, *Essays on Politics and Society*, 324–25.

61. *CW*, vol. 19, *Essays on Politics and Society*, 476.
62. *CW*, vol. 1, *Autobiography and Literary Essays*, 288–89.
63. *CW*, vol. 19, *Essays on Politics and Society*, 478.
64. Also reviewed in this essay were John Austin's *A Plea for the Constitution* (London: Murray, 1859) and James Lorimer's *Political Progress Not Necessarily Democratic; or, Relative Equality the True Foundation of Liberty* (London and Edinburgh: Williams and Norgate, 1857). For Mill's essay, see *CW*, vol. 19, *Essays on Politics and Society*, 341–70.
65. *CW*, vol. 15, *Later Letters*, 608.
66. *CW*, vol. 15, *Later Letters*, 653.
67. *CW*, vol. 19, *Essays on Politics and Society*, 343.
68. *CW*, vol. 19, *Essays on Politics and Society*, 358–67.
69. For an able analysis of the central role played by the principles of participation and competence in Mill's theory of representative government, see Thompson, *John Stuart Mill and Representative Government*.
70. *CW*, vol. 19, *Essays on Politics and Society*, 364.
71. *CW*, vol. 19, *Essays on Politics and Society*, 448–66.
72. *CW*, vol. 15, *Later Letters*, 653–54.
73. From the last letter written by Cobden before his death (to T.B. Potter, on 22 March 1865). See John Morley, *The Life of Richard Cobden*, 2 vols. (London: Macmillan, 1908), vol. 2, 493–94.
74. Mill had some misgivings about Hare's penchant for unsuitable quotations. Commenting to Helen Taylor in February 1860 on a recent paper by Hare in *Fraser's Magazine*, Mill observed: "On a subject which ought to be studiously presented in the most eminently practical light, his paper is overlaid with quotations of rhapsody from Carlyle & generalities from Maurice & Ruskin, as applicable to any subject as to this." *CW*, vol. 15, *Later Letters*, 667.
75. *CW*, vol. 19, *Essays on Politics and Society*, 370.
76. *CW*, vol. 19, *Essays on Politics and Society*, 457.
77. *CW*, vol. 19, *Essays on Politics and Society*, 473.
78. *CW*, vol. 19, *Essays on Politics and Society*, 477. Statements in a similar vein can be found at 443, 446, 478, 479, 508, 512, and 514.
79. *CW*, vol. 1, *Autobiography and Literary Essays*, 239.
80. *CW*, vol. 19, *Essays on Politics and Society*, 313, 314.
81. See his articles "The Contest in America" and "The Slave Power," in *CW*, vol. 21, *Essays on Equality, Law, and Education*, 125–42 and 143–64. The pro-Southern sympathies of most of the English governing class disgusted Mill. "I never before felt so keenly how little permanent improvement had reached the minds of our influential classes and of what small value were the liberal opinions they had got into the habit of professing." *CW*, vol. 1, *Autobiography and Literary Essays*, 267. Conversely, Mill found much to admire in the response of the Lancashire working classes to the 1862 cotton famine. On learning that there was to be a pro-Union demonstration in Manchester, Mill contrasted the "moral greatness" of "the suffering operatives of Lancashire" with "the mean feeling of so great a portion of the public on this momentous subject." *CW*, vol. 15, *Later Letters*, 813. Mill and John Bright fought on the same side in the quest to win over English public opinion, a prologue to their alliance on the issues of parliamentary reform and Jamaica in 1866 and 1867.
82. See Kinzer, Robson, and Robson, *A Moralist In and Out of Parliament*, 80–112.
83. For a different view of Palmerston's "liberalism," one that considers him a notably constructive political force, see E.D. Steele, *Palmerston and Liberalism 1855–1865* (Cambridge: Cambridge University Press, 1991).
84. For examples of the compliments Mill paid Gladstone during these years, see *CW*, vol. 28, *Public and Parliamentary Speeches*, 58, 89, 97–98, 323, 355–56, 360–61, 364–66.
85. *CW*, vol. 17, *Later Letters*, Appendix II, 2010–11.
86. See *CW*, vol. 28, *Public and Parliamentary Speeches*, 99–100, and *CW*, vol. 16, *Later Letters*, 1186.

NOTES

87. See *CW*, vol. 1, *Autobiography and Literary Essays*, 278, and *CW*, vol. 28, *Public and Parliamentary Speeches*, 103–04.
88. See *CW*, vol. 16, *Later Letters*, 1242–43, and *CW*, vol. 28, *Public and Parliamentary Speeches*, 103–04.
89. *CW*, vol. 16, *Later Letters*, 1191–92.
90. See John Vincent, *The Formation of the Liberal Party, 1857–1868* (London: Constable, 1966), 159.
91. For Mill's account of his connection with the Jamaica Committee, see *CW*, vol. 1, *Autobiography and Literary Essays*, 280–82. For studies of the controversy that pay considerable attention to Mill's role, see Bernard Semmel, *The Governor Eyre Controversy* (London: MacGibbon and Kee, 1962), especially chapter 3, and R.W. Kostal, *A Jurisprudence of Power: Victorian Empire and the Rule of Law* (Oxford: Oxford University Press, 2005). Kostal provides by far the most comprehensive treatment of the debate in his important work, which, among other things, criticizes aspects of the commentary on Mill's part given by Kinzer, Robson, and Robson in *A Moralist In and Out of Parliament*, chapter 6.
92. *CW*, vol. 16, *Later Letters*, 1126.
93. *Report of the Jamaica Royal Commission* 30 (1866): 489–531.
94. See Kinzer, Robson, and Robson, *A Moralist In and Out of Parliament*, 192–93.
95. For an account of the meeting that elected Mill chairman of the Jamaica Committee, see *The Times*, 10 July 1866.
96. *CW*, vol. 1, *Autobiography and Literary Essays*, 282.
97. *CW*, vol. 16, *Later Letters*, 1365.
98. The secondary literature on Mill and the Irish question is substantial: E.D. Steele, "J.S. Mill and the Irish Question: The Principles of Political Economy, 1848–1865," *Historical Journal* 13 (1970): 216–36; E.D. Steele, "J.S. Mill and the Irish Question: Reform and the Integrity of the Empire," *Historical Journal* 13 (1970): 419–50; R.N. Lebow, "J.S. Mill and the Irish Land Question," in *John Stuart Mill on Ireland*, ed. R.N. Lebow (Philadelphia: Institute for the Study of Human Issues, 1979), 3–22; Lynn Zastoupil, "Moral Government: J.S. Mill on Ireland"; T.A. Boylan and T.P. Foley, "John Elliot Cairnes, John Stuart Mill and Ireland: Some Problems for Political Economy," *Hermathena* (1983): 96–119; and Kinzer, *England's Disgrace?*
99. For the reaction to Mill's pamphlet, see Kinzer, *England's Disgrace?* 185–99.
100. *CW*, vol. 6, *Essays on England, Ireland, and the Empire*, 532.
101. Steele, "J.S. Mill and the Irish Question: Reform and the Integrity of the Empire," 437–46.
102. Zastoupil, "Moral Government: J.S. Mill on Ireland."
103. Kinzer, *England's Disgrace?* 171–85.
104. *CW*, vol. 1, *Autobiography and Literary Essays*, 278–79.

INDEX

Abbreviations used: TC = Thomas Carlyle; HT = Harriet Taylor; JM = James Mill; JSM = John Stuart Mill

Aberdeen Coalition, 162, 191
Adams, William Bridges, 121, 221 (n. 27)
American Civil War, 200, 203, 230 (n. 81)
Annual Register, 15
Anti-Jacobin Review, 31
Arabian Nights, 41
aristocracy
　ballot as weapon against, 148, 149, 153, 155
　contest between people and, 148, 149, 165
　government controlled by, 70
　inadequacies, 186
　JM and power of, 17–18
　liberality of better part of, 192
　means used to achieve power, 149
　parliamentary radicals fail to challenge, 190
　sinister interests of, 35, 181
　see also landed classes
association psychology, 3–4, 13, 29, 208 (n. 49)
Athens, 207 (n. 36), *see also* Greece
Austin, Charles, 3, 27–8, 35, 36, 46, 49, 67, 209 (n. 61)
Austin, John, 26, 28, 36, 48–9, 55–7, 65, 77, 117, 135, 186, 214 (n. 66)
Austin, Lucy, 27, 48
Austin, Sarah, 5, 26–7, 48–9, 55, 77, 88, 104, 105, 116, 117, 133–4, 135, 185, 209 (n. 60), 220 (n. 76)
Avignon, 209 (n. 76)

Baader, Franz von, 53
Bailey, Samuel, 148–9
Bain, Alexander, 11, 12, 36, 38, 47, 63, 210 (n. 93), 212 (n. 18)
ballot
　aristocratic power and, 148
　Bentham advocates, 17
　Chartists and, 153, 224 (n. 29)
　D.C. Moore's argument respecting, 160, 160–1
　electoral conditions and, 160–2
　HT and, 154–5
　Hume's "Little Charter" and, 168, 188
　impact of JSM on debate over, 159–60
　JM and, 146–7
　JSM advocates, 146–51, 163
　JSM on, 149, 150, 151, 152, 153, 154, 224 (n. 32)
　JSM opposes, 146, 156–9, 163, 193, 201
　JSM reconsiders, 155
　Philosophic Radicals and, 147, 148

INDEX

political comeback of, 155
scholarship on JSM and, 6–7
Whigs and, 149, 151
Ballot Society, 224 (n. 37)
Baring, Lady Harriet, 217–18 (n. 17)
Baring, William Bingham, 217 (n. 17)
Beales, Edmond, 227 (n. 42)
Bentham, George, 21, 22, 208 (n. 44) (n. 46)
Bentham, Jeremy
    Carlyle on, 116
    C.J. Hare's hostility to, 50–1
    death of, 77, 217 (n. 1)
    greatest happiness principle of, 18, 19
    human nature as viewed by, 16
    J. Austin and, 26
    JM and, 12, 15–16, 207 (n. 19), 218 (n. 27)
    JSM and, 18, 28–9, 30, 35, 35–6, 38, 130–2, 208 (n. 44), 211 (n. 96)
    motives categorized by, 207 (n. 20)
    and parliamentary reform, 17
    rewards and punishments in theory of, 17
    Sterling inveighs against, 62
    theory of government of, 16–17, 71
    Westminster Review and, 34, 77
    women's interests and, 87, 89, 218 (n. 27)
    writing methods of, 211 (n. 96)
    writings: *Introduction to the Principles of Morals and Legislation*, 87; *Plan of Parliamentary Reform*, 17; *Rationale of Judicial Evidence* (ed. JSM), 30, 35, 38, 211 (n. 96); *Traité de Législation* (ed. Dumont), 18, 28–9
Bentham, Maria Sophia (Lady Bentham), 21, 88
Bentham, Sir Samuel, 21, 26, 31, 208 (n. 44), 218 (n. 28)
Benthamism, 18, 27, 27–8, 28, 50, 130–2
Benthamites, 29, 30, 51, 66, 74, 77, 116, *see also* Philosophic Radicals
Berkeley, Francis Henry, 155, 159

Berkshire, 212 (n. 27)
Birch, A.H., 164
Blackstone, William, 198
blasphemy, 20
Bonn, 55, 56
botany, 21
Bourbons, 31
Boulogne, 220 (n. 4)
Bowring, John, 77
Bradlaugh, Charles, 21, 227 (n. 42)
Bright, John, 155, 156, 159, 179, 192, 197, 230 (n. 81)
British and Foreign School Society, 69
British and Foreign Unitarian Association, 77
Brodrick, G.C., 227 (n. 42)
Brougham, Henry, 69, 214 (n. 60), 216 (n. 109)
Buckinghamshire, 212 (n. 27)
Buller, Arthur, 113, 115
Buller, Charles (the elder), 113
Buller, Charles (the younger), 104–5, 113, 114, 115, 116, 133–4, 217–18 (n. 17)
Buller, Mrs. Charles (mother of the younger Charles Buller), 105, 113, 220 (n. 4)
Bulwer, Edward Lytton, 130, 131–2
Burke, Edmund, 72, 198
Buxton, Charles, 202
Byron, George Gordon (Lord), 51

Cairnes, J.E., 192
Calhoun, John C., 198
Cambridge Union, 27, 28, 50, 67, 105, 215 (n. 101)
Cambridge University, 27, 28, 36, 49, 49–50, 67, 114, 199, 213 (n. 38), 220 (n. 4), 227 (n. 42), *see also* Trinity College, Cambridge
Canada, 36, 152
Canning, George, 68, 70, 216 (n. 106)
Capaldi, Nicholas, 3, 5, 6
Carlisle, Janice, 3–4, 5, 210 (n. 85)

Carlyle, Alexander (brother of TC), 115
Carlyle, Jane Welsh, 46, 104, 113–14, 115, 117, 133, 134, 135, 138, 218 (n. 17), 220 (n. 3) (n. 4), 222 (n. 68)
Carlyle, John (brother of TC), 81, 115, 116, 118, 126, 133, 134, 135, 138, 140, 222 (n. 68)
Carlyle, Thomas
　articles on German literature of, 114, 213 (n. 51)
　attends University of Edinburgh, 113
　background of, 112–13
　on boundless capacity of man for loving, 121
　and changes in JSM's character, 132
　creed of, 128
　curious about JSM, 116
　on disastrous news brought by JSM, 138
　enthusiasm for Goethe, 114
　estrangement of JSM and, 142
　force of character of, 112
　fortunes of receive boost, 113
　and French Revolution, 136, 137, 137–8
　friendship of JSM and, 3, 6, 110, 112, 117, 117–18, 118, 122–3, 124–5, 126–7, 128–30, 133, 135–6, 142–5
　gains recognition, 142
　as "God-inspired man," 115
　HT fascinates, 222 (n. 88)
　on HT, 81, 134
　historical standing of, 6
　hostile to JM, 116
　introduced to HT, 134
　and Isabella Harriet Mill, 223 (n. 88)
　JSM on, 117, 118, 125, 128, 144
　on JSM, 46, 47, 116, 117, 118, 123, 124, 126–7, 128, 129–30, 134–5, 135–6, 138, 221 (n. 36), 223 (n. 97)
　JSM broaches question of differences with, 122–3
　JSM cancels visit to, 98, 100
　JSM contrasts own gifts with those of, 125
　on JSM and HT, 83–4, 104, 105
　JSM invokes, 94
　JSM and loss of manuscript of, 138–41
　JSM and mysticism of, 143–4
　JSM points out differences with, 128–9
　JSM reinforces impression of compatibility with, 129
　JSM reviews *French Revolution* of, 141–2
　JSM seeks to explain himself to, 124–5
　JSM thinks of as artist and poet, 125
　JSM urged to write history of French Revolution by, 137
　and JSM's appetite for "poetic feeling," 145
　and JSM's distinction between artist and logician, 119
　and JSM's exposure to German thought, 53
　and JSM's forthrightness, 128
　and JSM's ignorance of women, 84
　and JSM's "higher kind of sincerity," 133
　and JSM's involvement with Fox's circle, 134–5
　and JSM's involvement with HT, 104, 133–4
　on JSM's letters, 118
　and JSM's "Remarks on Bentham's Philosophy," 131–2
　and JSM's reply to "Negro Question" of, 203
　and JSM's *Spirit of the Age*, 115
　JSM's treatment of in *Autobiography*, 144
　JSM's writings show influence of, 121–2
　on John Taylor, 79
　Lady Harriet Baring and, 218 (n. 17)
　leaves Bullers, 114, 220 (n. 4)
　leaves Craigenputtock for London, 116
　longs for companionship, 116
　loss of faith and hardships of, 113

meets W.J. Fox, 134
on Madame Roland, 133
on the Mills' summer residence, 44
moves to Cheyne Row, 133
moves to Craigenputtock, 115
"negative part" of relations with JSM, 125
on poet as seer, 115
pursues Jane Welsh, 113–14, 220 (n. 3)
relates history to divine revelation, 136
responds to JSM's characterization of their differences,129–30
reputation of, 114
rising contempt of for HT, 134–5
Sterling's biographer, 51–2
as talker, 212 (n. 18)
tender feeling of for JSM, 135–6
tenor of JSM's letters to, 97
Thomas Hare quotes, 230 (n. 74)
watches JSM with deep interest, 124
welcomes JSM's greater candidness, 123
on W.J. Fox, 222 (n. 66)
writings: "Characteristics," 121; *French Revolution*, 138–43, 223 (n. 97); *Sartor Resartus*, 115, 116, 142, 143–4; "Signs of the Times," 115–16
Caroline, Queen, 68
Catherine the Great, 21
Catholic Association, 38
Catholic Church, 60
Catholic disabilities, 38
Catholic emancipation, 68, 185
Cecil, Lord Robert, 226 (n. 22)
Chadwick, Edwin, 174
Chapman, Henry S., 185, 186, 190
Chartism, 152, 153, 185, 187, 188, 229 (n. 38), *see also* Chartists
Chartists, 190, 224 (n. 29), *see also* Chartism
Choice of Hercules, 20
Christianity, 20, 20–1, 137
Church of England, 55, 70, 167, 177

civilization
and divorce, 96
JM on forming scale of, 65
JSM on natural tendency of modern, 198
JSM on Saint-Simonians and understanding of, 61
JSM on universal suffrage and state of, 187
law of development and course of, 59
sequential stages in development of, 65
sexual equality and progress of, 90
civil service, reform of, 185, 199
clerisy, 55, 227 (n. 51)
Clive, John, 209 (n. 61)
Cobden, Richard, 155, 162, 188, 197, 224 (n. 37), 230 (n. 73)
Cole, Henry, 47–8, 49, 51, 212 (n. 20) (n. 21)
Coleridge, Samuel Taylor, 50, 51, 52–5, 56, 63, 65, 177, 181, 214 (n. 54), 226 (n. 15)
Comte, Auguste, 58–9, 59–60, 214 (n. 70)
Condorcet, Marie Jean Antoine, 214 (n. 54)
Conservative party, 166–7, 167, 168, 170–1, 192, 226 (n. 15) (n. 17) (n. 22), 228 (n. 9), *see also* Conservatives, Tories
Conservatives, 152, 165, 170–1, 179–80, 183, 201, 202, 226 (n. 17), *see also* Conservative party, Tories
Conway, Moncure, 82, 83
Corn Laws, repeal of, 155, 199
Cornwall, 86, 212 (n. 27), 220 (n. 4)
Corrupt Practices Act (1854), 162
cotton famine, 230 (n. 81)
Craigenputtock, 98, 100, 104, 115, 116, 132
Cremer, W.R., 227 (n. 42)
Crimean War, 193
Crompton, J., 215 (n. 85)

Daily News, 154, 168, 169, 187, 187–90, 192
Daily Telegraph, 139
democracy
   TC's growing contempt for, 142
   English opponents of, 188
   Hare's scheme and true, 197
   and household suffrage, 180
   JSM on, 30, 149–50, 179–80, 187–8
   JSM's reservations regarding majoritarian, 198
   Tories see JSM as opponent of, 195
   Whigs view as threat, 70
Derby, Lord, 168, 173, 192, 226 (n. 15)
Disraeli, Benjamin, 168, 173, 179, 192, 226 (n. 15), 228 (n. 9)
divorce
   Fox supports, 106
   HT elicits JSM's thoughts on, 87
   HT on law of, 95
   HT's essay on marriage and, 94–7
   JSM's essay on marriage and, 89–94
   legal restrictions on, 89
   TC on Fox's circle and, 135
Doane, Richard, 209 (n. 70)
Don Quixote, 41
Dorking, 36, 48
Dumfriesshire, 113
Dumont, Etienne, 211 (n. 96)
Dunfermline, Lord, 162
Durham, Lord, 151–2, 182

East India College, 116
East India Company, 4, 32–5, 45, 113, 146, 191, 192, 229 (n. 50), *see also* India House
Ecclefechan, 113
Edgeworth, Maria, 41
Edinburgh, 113, 114, 115
Edinburgh Review, 12, 34, 66–70 *passim*, 88, 115, 115–16, 116, 121, 154, 166, 192, 213 (n. 51), 221 (n. 13)
Edinburgh, University of, 11, 113, 216 (n. 109)

education
   Coleridge's clerisy and, 55
   JM and, 13, 17, 69
   JSM on fear as element in, 25
   JSM on need for improvements in, 191
   JSM on plural voting and, 195, 198
   JSM on public spirit and, 158
   JSM's, 12–15, 18–20, 32, 41, 181, 207 (n. 13) (n. 19)
   party and civic, 177
   property and, 194, 195
   and suffrage, 190, 193, 194,
   Utilitarianism, Unitarianism, and, 77
   women and, 89, 90
Eichtal, Gustave d', 58, 59, 61, 62–3, 63, 74–5, 214 (n. 63) (n. 69), 216 (n. 125)
Empson, William, 116, 117, 221 (n. 13)
Encyclopædia Britannica, 66, 69
England
   American Civil War and, 230 (n. 81)
   blasphemy in, 20
   change in political and social organization of, 156–7
   distribution of education and property in, 194
   established professions in, 45
   fizzling out of Chartism in, 187
   Fox on middle classes in, 45
   and French Revolution, 30–1, 136, 137
   and Ireland, 203–4
   JM and reformation of society in, 65
   JM and JSM tour Southwest, 208 (n. 44)
   John Austin and JSM's view of society in, 56
   JSM on divergent paths of France and, 59
   JSM and electoral conditions in, 160–2
   JSM on French Revolution and, 136, 137
   JSM on ignorance in 136, 186

JSM on importance given to
    production in, 59
JSM on JM's contributions to, 23, 208
    (n. 49)
JSM on moral condition of, 162, 191,
    204, 214 (n. 63)
JSM on political condition of, 185,
    186, 190–1
JSM on problem of promulgating new
    doctrine in, 62
JSM returns from France to, 34
JSM's altered status in, 153, 184
JSM's intended audience in, 199
JSM's view of ballot and state of
    society in, 146
Macaulay on history of since
    Waterloo, 67
the Mills' social position in, 45
opponents of democracy in, 188
and radicalism of JM and JSM, 73
reaction against philosophy of 18th
    century in, 50
Roebuck's arrival in, 36
Saint-Simonian teachings and, 61
Sterling absent from, 213 (n. 43)
TC on ignorance of French
    Revolution in, 137
TC on JSM as mystic in, 122
England and the English, 130, 131–2
Enlightenment, 52, 53, 65, 77
Europe, 52, 57, 73, 78, 190, 203
Examiner, 81, 115, 117
Eyre, Edward John (Governor), 201–2

Falconer, Henrietta, 105
Faust, 114
Fawcett, Henry, 192, 229 (n. 54)
Fenianism, 203, 204
Fenton, Isabel (mother of JM), 11
Ferdinand VII (of Spain), 213 (n. 43)
Ferguson, Adam, 65
Fichte, Johann, 115
Finsbury, 77, 81, 86
Flower, Benjamin, 80

Flower, Eliza, 5, 80, 80–1, 81–3, 85,
    101–2, 104, 106, 107, 217 (n. 17)
Flower, Sarah, 80, 82, 83, 221 (n. 27)
Fonblanque, Albany, 115, 116
Forbes, William, 13, 19, 206 (n. 12)
Ford Abbey, 15, 208 (n. 44)
Fordyce, George, 21
Foreign Review and Continental Miscellany,
    213 (n. 51)
Forster, Edward, 41
Fox, Caroline, 46
Fox, Robert Barclay, 184, 185
Fox, W.J., 45, 77–83 *passim*, 86, 97,
    101–7 *passim*, 121, 134, 135,
    222 (n. 66)
Fox, Mrs. W.J., 80, 106, 107, 135
France
    Bullers consider residing in, 220 (n. 4)
    JM and, 30–1
    JM on restoration of Bourbons and, 31
    JSM on Christianity and, 137
    JSM on democracy in, 187–8
    JSM on divergent paths of England
        and, 59
    JSM and history of, 30
    JSM returns from, 25, 36
    JSM shares knowledge of with TC,
        142
    JSM's response to, 41
    JSM's sojourn in, 21–2, 34, 75,
        208 (n. 44)
    JSM's view of people of,
        209–10 (n. 76)
    Republican experiment in, 190
    Saint-Simonian teaching and, 61
    TC on hubbub in, 126
    see also French Revolution (1789,
        1830, 1848)
Fraser's Magazine, 142, 192
Freeman, E.A., 227 (n. 42)
free trade, 155, 199
French Revolution (1789), 30, 31, 41,
    57, 118, 136–8, 141, 142, 187
French Revolution (1830), 187

238 Index

French Revolution (1848), 186, 187, 229 (n. 36)
The Friend, 50

George IV, 68
Germany, 53, 56, 114, 115, 117
Girondists, 30, 31, 32, 136
Gladstone, W.E., 170–5 *passim*, 178, 192, 199, 200, 226 (n. 22), 230 (n. 84)
Glassman, P.J., 3
Goethe, Johann Wolfgang von, 53, 56, 114
Gompertz, Theodor, 171
Government
    Bentham's theory of, 16–17, 71
    competence and participation in JSM's theory of, 177, 230 (n. 69)
    Comte's inadequate conception of, 58
    democratic suffrage and conditions of, 190
    JM on scheme of relative to people's condition, 64
    JM's detestation of aristocratic, 31
    JM's theory of, 66, 71–2
    JSM on Hare's scheme and representative, 198
    JSM on Liberals and perfect model of, 167–8
    JSM on natural tendency of representative, 198
    JSM on prerequisite for honest, 158
    JSM on prospects for entering new era of, 150
    JSM on purposes of, 58
    JSM's view of Hare's scheme and representative, 198
    securities for good, 56, 88
Graham, George John, 36–7, 47, 49, 79, 210 (n. 91), 213 (n. 29)
Graham, Sir James, 162
Grant, Horace, 47, 48, 49
Great Exhibition (1851), 212 (n. 20)
Greece, ancient, 31–2, 143, *see also* Athens

Greville, Charles, 46–7, 212 (n. 17), 217 (n. 17), 217–18 (n. 17)
Grey, Lord, 69, 216 (n. 106)
Grote, George, 47, 48, 49, 151, 156, 182, 209 (n. 71), 214 (n. 60)
Grote, Harriet, 47, 48, 49, 88, 209 (n. 71)
Guizot, François, 198

Halévy, Elie, 16
Hamann, Johann Georg, 53
Hamburger, Joseph, 146, 148, 165, 181
Hamilton, Sir William, 200
Hampshire, 48, 84
Hare, Julius Charles, 50–1, 51
Hare, Thomas, 164, 167, 169, 192, 195, 198, 229 (n. 54), 230 (n. 74)
Hartley, David, 13, 29
Hayek, F.A., 5
Helvetius, Claude Adrien, 16
Herbert, Auberon, 227 (n. 42)
Herder, Johann Gottfried, 53
Hobbes, Thomas, 72
Holyoake, G.J., 160
Hooker, William Jackson, 212 (n. 23)
household suffrage, *see* suffrage
Howell, George, 227 (n. 42)
Hume, David, 16
Hume, Joseph, 154, 168, 188, 190, 192, 214 (n. 60)
Huskisson, William, 199
Hyde Park riots, 201

India, 33, 35, 113, 229 (n. 50)
India Act (1858), 229 (n. 50)
India House, 4, 12, 32–5, 36, 38, 40, 43–5 *passim*, 48, 105, 110, 174, 210 (n. 84), 212 (n. 20), 219 (n. 63), *see also* East India Company
Ireland, 152, 161, 173, 184, 185, 186, 201, 203–4, 231 (n. 98)
Irving, Edward, 113, 114, 220 (n. 3)
Isle of Wight, 48, 78, 84

INDEX

Jacobi, Heinrich, 53
Jacobins, 30, 67, 69
Jacobs, Jo Ellen, 5
Jamaica, 173, 201–2, 230 (n. 81)
Jamaica Committee, 202, 231 (n. 91) (n. 95)
Jeffrey, Francis, 69, 166

Kamm, Josephine, 5
Kant, Immanuel, 53, 213 (n. 50)
Kaplan, Fred, 6
Kensington, 44, 133, 139
Kent, 139
Kent Terrace (Regent's Park), 86, 97, 98, 103, 139
Keston Heath, 139
Kew Gardens, 212 (n. 23)
Kilmarnock, 174
Kingston, 139
Kirkcaldy, 113
Kostal, R.W., 231 (n. 91)

laboring classes, *see* working classes
Lancashire, 230 (n. 81)
landed classes
    Coleridge on role of, 54
    Coleridge's view of, 54, 55
    control political system, 17
    electoral influence of, 148, 161, 194
    JSM on ballot and power of, 156–7
    party conflict and, 148
    see also aristocracy
Latham, William, 159
law
    Bentham as philosopher and reformer of, 15, 17, 131
    Brougham and reform of, 216 (n. 109)
    and equal rights for women, 88
    HT and injustice inflicted on women by, 93
    Jamaica question and martial, 202
    Jamaica question and rule of, 202
    JSM reads Roman, 26, 27
    as political instrument of aristocracy, 70
    as possible career for JSM, 26, 32–3, 36
    profession of, 45
Lewes, George Henry, 144
Lewis, George Cornewall, 157, 196
Liberal party
    Gladstone's journey toward, 192
    JSM and formation of advanced, 172–3, 175, 200
    JSM and Gladstonian, 177–8
    JSM on merits of, 170–1
    and JSM's conception of his role in Parliament, 172
    and JSM's conception of two-party system, 166
    JSM's ideal of, 177–8
    JSM's view of condition of, 168, 171–2, 176
    JSM's view of Gladstone and, 170, 174
    JSM's view of purpose of, 167
    JSM's view of working classes and, 175–6
    narrowing of gap between Conservative party and, 226 (n. 22)
    see also Liberals
Liberals, 152, 159, 165, 167–8, 169, 170–1, 174–5, *see also* Liberal Party
"Little Charter," 154, 168, 188, 190, 192
Liverpool, Lord, 67, 68, 199
Locke, John, 13, 29
London, 11, 26, 30, 36, 44, 58, 77, 78, 80, 86, 97, 103, 104, 105, 113, 115, 116, 117, 126, 133, 134, 135, 213 (n. 43), 220 (n.4) (n. 76)
London Debating Society, 29, 46, 47, 50, 51, 53, 58, 62–3, 63, 105, 214 (n. 79), 216 (n. 125)
London, University of, 55, 69, 116, 214 (n. 60), 215 (n. 85), *see also* University College, London
London and Westminster Review, 141, 149, 180–1, 182, 183, 221 (n. 34), *see also Westminster Review*
Lovett, William, 187

Lowe, Robert, 159, 179
Lushington, Godfrey, 227 (n. 42)

Macaulay, Hannah, 67
Macaulay, T.B., 27, 28, 66, 67, 68–9, 69, 70, 71–2, 72, 73, 209 (n. 61), 212 (n. 18), 215 (n. 101), 216 (n. 110)
Manchester, 68, 230 (n. 81)
Marmontel, Jean François, 39, 40
marriage
  Fox's views on, 80, 135
  HT's essay on, 94–7
  JSM and HT find fault with, 87
  JSM's essay on, 89–94
  TC on Fox's circle and, 135
Martineau, Harriet, 5, 79, 80, 83, 105–6
Maurice, F.D., 49, 49–52 *passim*, 230 (n. 74)
Mazlish, Bruce, 3, 206 (n. 5)
Mazzini, Giuseppe, 218 (n. 17)
Melbourne, Lord, 149, 150
mental crisis, JSM's, 29, 37–42, 182
Methodism, 37
middle classes
  competitive universe of, 45
  conservatism of some members of, 188
  Fox on, 45
  and Hare's scheme of personal representation, 197
  independence of from upper classes, 157
  and intellectual elite, 227 (n. 51)
  JM and lodging of power in, 71
  and JM's position on suffrage, 215 (n. 99)
  JSM on ballot and, 153
  JSM on unfixedness in, 45
  JSM views as "natural Radicals," 152
  Macaulay, JM, and enfranchisement of, 69
  political cooperation between working classes and, 153
  and radicalization of Liberal party, 176
  and universal suffrage, 152
Mill, Clara (sister of JSM), 218–19 (n. 28)

Mill, Harriet Burrow (mother of JSM), 5, 11–12, 46, 47, 63, 84, 87, 88, 206 (n. 10)
Mill, Harriet Isabella (sister of JSM), 140, 223 (n. 88)
Mill, James
  and association psychology, 29
  and Austins, 27
  Bain on tense family atmosphere created by, 63
  and ballot, 147, 156, 160
  on ballot, 146–7
  and Bentham, 12, 15–16, 207 (n. 19), 218 (n. 27)
  and Brougham, 216 (n. 109)
  childhood and education of, 11, 207 (n. 35)
  collections of essays by, 215 (n. 95)
  and daughters' education, 87–8
  and democratic suffrage, 215 (n. 99)
  and *Edinburgh Review*, 12, 69–70
  and *Elements of Political Economy*, 25–6
  epistemology of, 73
  fear inspired by, 23–4, 25
  and fear of revolution, 216 (n. 108)
  and France, 30–1
  George Bentham on, 208 (n. 44)
  and guardianship of JSM, 207 (n. 19)
  habits of, 138
  handsomely provides for family, 43–4
  at India House, 32, 33, 45
  JSM admonished by, 21–2
  JSM on ballot and, 156
  JSM on contributions of, 23, 208 (n. 49)
  JSM on education given by, 14
  JSM and final illness and death of, 182
  JSM on moral convictions of, 19
  JSM and pervasive presence of, 22
  JSM propagates doctrines of, 35–6
  JSM quarrels with, 36–7, 210 (n. 93)
  and JSM's anxiety, 24–5
  and JSM's *Autobiography*, 9–10
  and JSM's career path, 32–3
  and JSM's development, 2, 3, 10, 64–5

JSM's distant relations with, 46, 63–4, 125
and JSM's education, 12–15, 18–20
on JSM's education, 13–14
JSM's fear of, 22–3
and JSM's legal studies, 28
and JSM's mental crisis, 37–42 *passim*
and JSM's performance at India House, 35
and JSM's program of internal culture, 63
and JSM's "Remarks on Bentham's Philosophy," 131, 132
Macaulay attacks, 66, 67, 71–2
as man of action and practical reformer, 4
marriage of, 11–12
Moore's view of, 160
moral convictions of, 19–20, 143, 207 (n. 35)
Peacock as colleague of, 47
Place on educational method of, 15
political reasoning of, 71–3
and reformation of English society, 65
Roebuck on JSM's quarrel with, 210 (n. 93)
and Samuel Bentham, 208 (n. 44)
on scheme of government relative to people's condition, 64
struggles of, 12
on superior spirits and social progress, 65
TC and, 112–13, 115, 116
theory of government of, 66, 71–2
Tooke and influence of, 49
and University of London, 55
and *Westminster Review*, 34
and Whigs, 69
and Wilhelmina Stuart, 11
and women's character, 88
and women's interests, 71
and women's suffrage, 88–9
writings: *Analysis of the Phenomena of the Human Mind*, 29, 208 (n. 49); *Elements of Political Economy*, 25–6;
Essay on Government, 66, 69, 70–1, 72, 73, 87, 215 (n. 97); *History of British India*, 11, 12, 23, 47, 64–5, 69, 146–7, 156, 215 (n. 85)

Mill, James Bentham (brother of JSM), 46

Mill, John Stuart, writings
Autobiography, 8–9, 13, 21, 24, 37, 39, 49, 53, 56, 64, 71, 75, 76, 81, 87, 88–9, 110, 130, 144, 154, 162, 163, 172, 179, 180, 182–3, 186, 191, 193, 195, 202, 204, 206 (n. 4), 209 (n. 73), 211 (n. 96), 213 (n. 50), 214 (n. 79), 217 (n. 10)
"Bain's Psychology," 192
"Bentham," 181
"Carlyle's French Revolution," 141–2, 142–3
"Civilization," 148
"Coleridge," 181
Considerations on Representative Government, 164–5, 170, 176–7, 187, 193, 194–5, 195, 195–6, 197, 198, 199
"The Contest in America," 200, 203
"De Tocqueville on Democracy in America [I]," 148
"De Tocqueville on Democracy in America [II]," 148, 183, 184, 185
Dissertations and Discussions, 192, 197, 229 (n. 53)
England and Ireland, 203–4
Essays on Some Unsettled Questions of Political Economy, 221 (n. 34)
An Examination of Sir William Hamilton's Philosophy, 200
"A Few Words on Non-Intervention," 192
"Flower's Songs of the Seasons," 81–2
"French Affairs," 187
"Mignet's French Revolution," 136
"Milnes's Poems," 144
"The Negro Question," 203
"Notes on the Newspapers," 166

Mill, John Stuart, writings—*continued*
  "On Genius," 31, 121, 144
  "On Liberty," 122, 185, 191, 192, 200, 229 (n. 53)
  "On Marriage," 89–94, 125
  "On the Definition of Political Economy," 221 (n. 34)
  Principles of Political Economy, 8, 33, 155, 156, 184, 185
  "Rationale of Representation," 72, 148–9
  "Recent Writers on Reform," 192, 193, 196, 199
  "Reform of the Civil Service," 185, 191
  "Remarks on Bentham's Philosophy," 130–2
  "Reorganization of the Reform Party," 152–3, 176, 183, 188, 226 (n. 15)
  "Scott's Life of Napoleon," 136
  "The Slave Power," 200, 203
  Spirit of the Age, 65, 115, 116, 117, 214 (n. 66)
  Subjection of Women, 87
  System of Logic, 8, 33, 72, 126, 154, 155, 156, 182, 184, 228 (n. 11)
  "Tennyson's Poems," 144
  Thoughts on Parliamentary Reform, 146, 154, 155, 156–9, 159–60, 161, 162, 187, 191, 192, 193, 194–5, 196, 199
  "Thoughts on Poetry and Its Varieties," 144
  Three Essays on Religion, 21
  "Vindication of the French Revolution of February 1848," 187
  "Writings of Alfred de Vigny," 144
  "Writings of Junius Redivivus," 121
Mill, Wilhelmina (sister of JSM), 218–19 (n. 28)
Millar, John, 65
Milne, James (father of JM), 11
Milnes, Richard Monckton, 213 (n. 38)
Milton, John, 15

Monthly Repository, 77, 79, 80, 106, 121, 122, 135
Montpellier, University of, 22
Montrose Academy, 11
Moore, D.C., 160–1
Morant Bay, 202
Morning Chronicle, 185, 186, 203
music, 41, 48, 79, 82
mysticism, 130, 143–4

Napier, Macvey, 69, 184, 221 (n. 13)
Napoleon, 30, 211 (n. 2)
Napoleon III, 208
National Association for Promoting the Political and Social Improvement of the People, 187
Navigation Acts, 199
Neff, Emery, 6
Nichol, John Pringle, 75, 132, 216 (n. 129)
Northern Star, 224 (n. 29)
Norton, Charles Eliot, 83, 139, 218 (n. 18)
Norwich, 26, 27, 215 (n. 85)
Nossiter, T.J., 162

O'Connell, Daniel, 38
Odger, George, 176
On the Constitution of the Church and the State, 54, 55
open voting, *see* voting, open
Oxfordshire, 212 (n. 27)
Oxford University, 199, 227 (n. 42)

Packe, Michael St. John, 3, 5, 6, 218 (n. 17)
Palmerston, Lord, 162, 165, 168, 170, 172, 176, 192, 193, 200, 201, 226 (n. 22), 230 (n. 83), *see also* Palmerstonian ascendancy
Palmerstonian ascendancy, 165, 169, 176–7, *see also* Palmerston, Lord
Pankhurst, Richard, 6
Panopticon, 15, 17
Paris, 97, 98, 100, 101, 103, 104, 106, 107, 126, 132, 219 (n. 63)
Parks Bill (1867), 201

Parliamentary History and Review, 29–30, 38, 38–9
parliamentary reform
  Bentham and, 17
  Bright and, 192, 230 (n. 81)
  fails to materialize in 1859–60, 200
  Gladstone and, 173
  Hare's proposals and, 195–8
  JM and, 66, 69, 72
  JSM and, 154, 155, 171, 180, 189, 189–90, 192, 199, 230 (n. 81)
  placed on back burner, 193
  radicals and, 188
  revival of interest in, 168, 170
  Tory ministry moves on, 193
  see also suffrage
party
  and antagonism in JSM's thought, 165
  and civic education, 177
  central to JSM's political activity in 1830s, 165–6
  and Hare's scheme of personal representation, 169–70
  JSM on his unfitness for leadership of, 182
  JSM's campaign to create viable radical, 149–50, 181
  JSM's ideal of, 166, 177–8
  and JSM's parliamentary career, 170–6, 201
  Philosophic Radicals and, 148, 165–6
  scholarship on JSM and, 6–7
  treatment of in *Considerations on Representative Government*, 164–5
  in Victorian political life, 164
  see also Conservative Party, Liberal Party
Peacock, Thomas Love, 36, 47
Peel, Sir Robert, 44, 149, 151, 185, 199, 226 (n. 15)
Peterloo, 68
Philosophic Radicals, 36, 50, 146, 147, 148, 155, 165–6, 181, 181–2, *see also* radicals, Utilitarianism, Utilitarians
Pitt, William (the younger), 44

Place, Francis, 11, 15, 31, 206 (n. 10), 214 (n. 60)
Plummer, John, 171, 175
plural voting, *see* voting, plural
poetry, 79, 115, 125, 136, 136–7, 144
political economy, 14, 18, 25–6, 35, 75
Presbyterianism, 19
Producteur, 59, 61
progress
  JM and social, 17–18, 64, 65, 216 (n. 109)
  JSM on halting of, 191
  JSM on institutions and stages of, 64
  JSM on Liberal party and cause of, 174
  JSM and prospect of political, 201
  politics as instrument for accelerating, 190
  radical party and promotion of, 181
Puckler-Muskau, Prince Hermann, 220 (n. 76)

Quarterly Review, 34

radicalism
  Bentham's turn toward political, 17
  Chartism and middle-class, 229 (n. 38)
  cooperation between middle and working classes essential to, 153
  ebb and flow of JSM's, 7
  Fox's, 78
  JSM on essence of, 152
  JSM and fortunes of, 184
  JSM and HT cultivate lofty, 191–2
  JSM and orthodox democratic, 190
  JSM on Peel and, 226 (n. 15)
  JSM on political action and, 183
  JSM's apprenticeship in Utilitarian, 40
  JSM's continued political, 187
  and "Reorganization of the Reform Party," 183
  respectable London, 77
  theory, practice, and JSM's, 204
  see also Philosophic Radicals, radicals

radicals
  Bright urges to support Palmerston and Russell, 192
  Buller joins ranks of, 105
  Chartist suspicion of middle-class, 153, 224 (n. 29)
  differences between JSM and, 156, 193, 195, 197, 198
  diminished utility of ballot to, 153
  dispirited, 151
  fare poorly in 1868 election, 174
  feebleness of parliamentary, 150, 190
  in France and Revolution of 1848, 187
  Grotes open home to, 48
  JSM on Gladstone's ministry and, 174–5
  JSM on need for effective leadership of, 150
  JSM on political action and, 183
  JSM on prospects of, 151
  JSM seeks to instruct, 188–9
  JSM works to foster cause of, 149
  JSM on Whigs and, 151, 176
  JSM's contempt for most, 156
  JSM's depiction of "natural," 152, 183
  and JSM's plea for large dose of parliamentary reform, 190
  JSM's view of political order and potential influence of, 181
  platform of in 1840s, 187
  and reshaping of political alignments, 148
  and revival of reform cause, 188
  see also Philosophic Radicals, radicalism
Reform Act (1832), 146, 150, 161, 181, 185
Reform Act (1867), 161, 175, 180
Reformation (Protestant), 20
Reform Bill (1831), 150
Reform Bill (1854), 191
Reform Bill (1859), 192
Reform Bill (1866), 173, 175, 202
Reform League, 200–1, 201
representation
  of educated, 196–7
  JM's treatment of, 66
  JSM and Hare's plan of personal, 164, 169–70, 195–8
  of working classes, 175, 177, 189–90
  see also, government, parliamentary reform, suffrage
Ricardo, David, 18, 24, 64
Richter, Jean Paul, 53
Robespierre, Maximilien, 136
Robinson Crusoe, 41
Robson, Ann P., 215 (n. 85)
Robson, John M., 1, 6, 154, 165, 224 (n. 32)
Roebuck, John Arthur, 36–7, 47, 49, 51, 67, 79, 84, 105, 105–6, 210 (n. 90) (n. 93)
Roland, Mme, 133
Romilly, Henry, 159
Romilly, John, 28, 46
Rose, Phyllis, 5
Roundell, Charles, 227 (n. 42)
Ruskin, John, 230 (n. 74)
Russell, Lord John, 150, 151, 162, 168, 170, 186, 188, 192, 200
Russia, 21

Saint-Simon, Claude-Henri, 57–8, 59, *see also* Saint-Simonians
Saint-Simonians, 57–63, 65, 126, 181, 216 (n. 125)
Say, J.B., 75
Say, Mme, 75
Schleiermacher, Friedrich, 53
Scotland, 10, 11, 65, 98, 112–13, 117–18, 126
Scott, Sir Walter, 136
Second Lay Sermon, 54–5
secret ballot, *see* ballot
secret voting, *see* ballot
sexual equality, 87–94
Sidgwick, Henry, 164
sinister interests, 35, 60, 70, 147, 181
Smith, Adam, 65
Smith, Sydney, 212 (n. 18)

Smith, Thomas Southwood, 77, 217 (n. 1)
Smollett, Tobias, 41
Society for the Diffusion of Useful Knowledge, 69
Society of Students of Mental Philosophy, 29
Socrates, 19, 20, 207 (n. 36)
Solly, Henry, 46, 216 (n. 129)
Songs of the Seasons, 81–2
Southern, Henry, 77
Southey, Robert, 47
South Place Chapel, 77, 80, 82, 83, 107, 135
Spain, 213 (n. 43)
Stanley, Lord, 226 (n. 22)
Steele, E.D., 204, 230 (n. 83)
Stephen, Leslie, 38, 215 (n. 97), 227 (n. 42)
Sterling, John, 49, 50, 51–2, 53, 55, 62–3, 73, 74, 82, 117–18, 182, 213 (n. 38) (n. 43), 214 (n. 79), 218 (n. 17)
Stillinger, Jack, 206 (n. 4)
Strutt, Edward, 28, 46
Stuart, Lady Jane, 11
Stuart, Sir John, 11
Stuart, Wilhelmina, 11, 12, 13, 206 (n. 12)
suffrage
  Bentham advocates universal, 17
  and *Considerations on Representative Government*, 198
  and educational qualification, 190, 193
  Hare's conservative views on, 198
  household, 152, 154, 168, 179, 180, 186–7, 188
  JM and question of, 66, 71, 88–9, 215 (n. 99)
  JSM dissents from Reform's League's position on, 201
  JSM on effect of extending, 198
  mode of voting and extent of, 153, 158–9
  and plural voting, 194, 195
  universal, 17, 152–3, 153, 158–9, 176, 186, 187, 190, 195, 196–7, 198, 215 (n. 99)
  women's, 66, 87, 88–9, 173
  see also parliamentary reform, representation
Surrey, 44, 46, 212 (n. 27)
Sussex, 48, 84, 212 (n. 27)

"taxes on knowledge," 199
Taylor, Harriet
  aspirations of, 79
  Bain on Fox and, 78–9
  and ballot, 154–5
  burial place of, 209 (n. 76)
  captivating aura cast by, 78, 81
  Carlisle's treatment of JSM's friendship with, 4
  the Carlyles' fascination with, 222 (n. 68)
  death of, 192, 199
  description of, 78
  difficulties experienced by JSM and, 103–4
  dilemma of, 102
  effects of aspirations and aversions of, 97
  Eliza Flower and, 80–1, 82–3
  emotional aggressiveness of, 100
  emotional resilience and self-fashioning of, 110
  first meeting between JSM and, 83, 84, 217 (n. 10)
  force of character of, 84, 112
  Fox and triangular compact contrived by, 106
  Fox's role in JSM's involvement with, 78–9, 79–80
  on happiness as word without meaning, 104
  height of, 217 (n. 2)
  influence of on JSM's emotional life, 110, 112
  and institution of marriage, 87
  John Taylor and, 78, 79, 97, 98, 101, 102, 103
  JSM acknowledges superior nature of, 94

Taylor, Harriet—*continued*
  JSM on belief of regarding John Taylor, 103
  JSM on doubts of, 101
  JSM enters orbit of, 77
  JSM on exemplary status of, 93–4
  JSM exempts from precepts regulating conduct of ordinary people, 93
  JSM on happiness of in Paris, 104
  JSM on impact of Paris episode on his relations with, 101–2
  JSM on initial interest of in him, 87, 89
  JSM introduces Carlyles to, 134
  JSM on John Taylor's admirable treatment of, 102
  JSM offends, 109
  JSM on poetic elements in character of, 111
  and JSM seek company of Fox and Flower, 104
  JSM on state of his affairs with, 107
  JSM on trial separation of John Taylor and, 97
  JSM upset by, 107–8
  JSM upsets, 107–8
  JSM visits at Kent Terrace, 86
  and JSM's advantages, 76
  and JSM's appetite for "poetic feeling," 144–5
  JSM's assignations with, 86–7
  on JSM's autobiographical project, 9
  and JSM's *Autobiography*, 8, 9, 10, 22, 209 (n. 73), 220 (n. 90)
  and JSM's avowal of change in his character, 132–3
  JSM's circumspect exchanges with, 125
  and JSM's criticism of TC's characterization of Mme Roland, 133
  and JSM's emotional reserve, 98
  and JSM's intellectual development, 110–11
  and JSM's perplexity, 118, 121
  on JSM's self-doubt, 99–100
  letter attesting to JSM's romantic friendship with, 84–5
  lofty radicalism of, 191
  on marriage and divorce, 94–7
  married life of JSM and, 191
  modern responses to JSM's friendship with, 2, 5–6
  Paris rendezvous of JSM and, 101–2
  passion for JSM of, 86, 102
  and plural voting, 193, 194
  political independence of JSM and, 155–6
  regularizing of JSM's friendship with, 182
  regulates relationship with JSM, 107
  Roebuck cautions JSM regarding involvement with, 105–6
  Roebuck on JSM's involvement with, 84, 105–6
  Sarah Austin reacts to JSM's involvement with, 133–4, 209 (n. 60), 220 (n. 76)
  self-confidence of, 100–1
  self-esteem of, 100
  status of JSM's relations with, 98–9
  TC on, 81, 83–4, 134, 135
  TC learns of JSM's involvement with, 133–4
  and TC's lost manuscript, 138, 139–41
  TC's rising contempt for, 134–5
  unwilling to sever ties with JSM, 97
Taylor, Helen, 5, 192, 229 (n. 50), 230 (n. 74)
Taylor, Henry, 46, 46–7, 47, 49, 212 (n. 17)
Taylor, Herbert (son of John Taylor and HT), 78
Taylor, John, 77, 78, 79, 81, 83, 86, 97, 98, 101, 102, 103, 106, 108–9, 134, 135, 139
Thomas, William, 207 (n. 35)
Thompson, Dennis, 164, 177
Thomson, Dr. Thomas, 32

Index

Thornton, W.T., 174
Times, The, 140, 159, 180, 201
Tocqueville, Alexis de, 148, 150, 183, 184, 185
Tooke, Thomas, 49, 75
Tooke, William Eyton, 49, 50, 74–6, 216 (n. 125) (n. 129)
Tories, 50, 70, 148, 149, 150, 151, 165, 166, 167, 169, 176, 180, 181, 187, 192, 193, 195, 198, 224 (n. 13), *see also* Conservative party, Conservatives
Toulouse, 21
triennial parliaments, 154, 168, 188, *see also* parliamentary reform
Trinity College, Cambridge, 50, 114, *see also* Cambridge University

Unitarianism, 77, 80
universal suffrage, *see* suffrage
University College, London, 46, *see also* London, University of
Urbinati, Nadia, 164
Utilitarianism, 77, 116, 127, 129, 130, 216 (n. 110), *see also* Philosophic Radicals, Utilitarians, Utilitarian Society
Utilitarians, 66–70 *passim*, 120, 129, 131, *see also* Philosophic Radicals, Utilitarianism, Utilitarian Society
Utilitarian Society, 29, 36

Victoria and Albert Museum, 212 (n. 20)
Villiers, Charles, 28, 46
Villiers, Hyde, 28
Virginia, 82
Voting
  open, 154, 156–9, 159–60, 160, 163, 190, 193, 224 (n. 13)
  plural, 193, 193–4, 194–5, 195, 196, 198
  secret, *see* ballot
  see also parliamentary reform, suffrage

Wales, 86, 161
Walpole, Spencer, 201
Walton-on-Thames, 139
Waterloo, 67
Welsh, Jane Baillie, *see* Carlyle, Jane Welsh
Welsh, John, 114
Welsh, Mrs. John, 114
West, A.S., 215 (n. 85)
Westminster, 15, 20, 170, 171, 173, 183, 201, 204
Westminster Review, 30, 34, 49, 69–70, 77, 136, 146, 147, 166, *see also* London and Westminster Review
Whewell, William, 228 (n. 11)
Whigs, 68–9, 69, 70, 148, 149, 150, 151, 165–6, 166, 173, 176, 181, 187, 188
Wilhelm Meister, 114
Women
  Bentham and interests of, 87
  education and emancipation of, 90
  Fox supports equal rights for, 78
  HT's treatment of condition of, 94–6
  JM and interests of, 66, 71, 87
  JSM on beautifying of life by, 90–1
  JSM on interest of, 89
  JSM on suffrage and, 88, 89
  JSM supports equal rights for, 88–9
  and JSM's essay on marriage and divorce, 89–94
  and JSM's plural voting scheme, 194
  and suffrage, 66, 87, 88, 89, 173
Wordsworth, William, 41, 48, 50, 51, 53, 63, 115, 181, 213 (n. 38)
working classes
  cotton famine and Lancashire, 230 (n. 81)
  educational qualification and electoral power of, 193
  and employment of women, 90
  and function of intellectual elite, 227 (n. 51)
  growing independence of, 157
  JSM on political strength of, 175

working classes—*continued*
  JSM reaches out to, 201
  JSM on suffrage and, 175, 187, 190, 198
  JSM views as "natural Radicals," 152
  and JSM's readership, 199
  and Liberal party, 175, 176, 177, 200
  need for cooperation between middle and, 153
  need for enfranchisement of, 175, 186–7, 189–90
  and Second Reform Act, 175
  and universal suffrage, 152–3

Xenophon, 19

Yorkshire, 48
Young, George, 227 (n. 42)

Zastoupil, Lynn, 204